FLOODED

NATURE, SOCIETY, AND CULTURE
Scott Frickel, Series Editor

A sophisticated and wide-ranging sociological literature analyzing nature-society-culture interactions has blossomed in recent decades. This book series provides a platform for showcasing the best of that scholarship: carefully crafted empirical studies of socio-environmental change and the effects such change has on ecosystems, social institutions, historical processes, and cultural practices.

The series aims for topical and theoretical breadth. Anchored in sociological analyses of the environment, Nature, Society, and Culture is home to studies employing a range of disciplinary and interdisciplinary perspectives and investigating the pressing socio-environmental questions of our time—from environmental inequality and risk, to the science and politics of climate change and serial disaster, to the environmental causes and consequences of urbanization and war making, and beyond.

For a list of all the titles in the series, please see the last page of the book.

FLOODED

Development, Democracy, and Brazil's Belo Monte Dam

PETER TAYLOR KLEIN

RUTGERS UNIVERSITY PRESS
New Brunswick, Camden, and Newark, New Jersey, and London

Library of Congress Cataloging-in-Publication Data

Names: Klein, Peter Taylor, author.
Title: Flooded : development, democracy, and Brazil's Belo Monte Dam / Peter Taylor Klein.
Description: New Brunswick : Rutgers University Press, [2022] | Includes bibliographical references and index.
Identifiers: LCCN 2021041973 | ISBN 9781978826137 (cloth) | ISBN 9781978826120 (paperback) | ISBN 9781978826144 (epub) | ISBN 9781978826151 (mobi) | ISBN 9781978826168 (pdf)
Subjects: LCSH: Belo Monte (Power Plant)—Social aspects. | Dams—Social aspects—Brazil—Pará (State) | Hydroelectric power plants—Social aspects—Brazil—Pará (State) | Sustainable development—Brazil—Pará (State) | Xingu River (Brazil)—Environmental conditions. | Pará (Brazil : State)—Social conditions. | Environmental policy—Brazil.
Classification: LCC TK1442.B45 K55 2022 | DDC 621.31/2134098115—dc23/eng/20220124
LC record available at https://lccn.loc.gov/2021041973

A British Cataloging-in-Publication record for this book is available from the British Library.

Copyright © 2022 by Peter Taylor Klein
All rights reserved

No part of this book may be reproduced or utilized in any form or by any means, electronic or mechanical, or by any information storage and retrieval system, without written permission from the publisher. Please contact Rutgers University Press, 106 Somerset Street, New Brunswick, NJ 08901. The only exception to this prohibition is "fair use" as defined by U.S. copyright law.

References to internet websites (URLs) were accurate at the time of writing. Neither the author nor Rutgers University Press is responsible for URLs that may have expired or changed since the manuscript was prepared.

∞ The paper used in this publication meets the requirements of the American National Standard for Information Sciences—Permanence of Paper for Printed Library Materials, ANSI Z39.48-1992.

www.rutgersuniversitypress.org

Manufactured in the United States of America

For Stephanie

CONTENTS

	Prologue	1
	Introduction	11
	PART I: HYDROPOWER, RESISTANCE, AND THE STATE	
1	Dams and Development	27
2	Booms, Busts, and Collective Mobilization along the Transamazon	46
3	Democratic Developmentalism	64
	PART II: AN ETHNOGRAPHY OF DAM BUILDING	
4	The Living Process	89
5	The Fight for Recognition	113
6	The Law, Activism, and Legitimacy	139
	Conclusion	166
	Acknowledgments	183
	List of Abbreviations	187
	Notes	189
	Index	215

FLOODED

PROLOGUE

I was in the back seat of the public defenders' pickup truck, speeding down the newly paved Transamazon Highway. It was October 2012, over a year after construction began on Belo Monte, which would become one of the five largest hydroelectric facilities in the world. We were heading from the city of Altamira, the closest urban center, toward one of the dam's construction sites. At the time, I had spent twelve months in Brazil during the previous three years. I was in the midst of my research on how dam construction was transforming the region's social, political, and environmental landscapes, and was particularly interested in the challenges and opportunities people faced in making claims for their livelihoods in this context.

Andreia Macedo Barreto, one of the two public defenders in the truck, asked me to put on music to soothe her as she drove. Andreia was a committed, fierce supporter of dam-affected communities. Born and raised in Belém, the capital of the forested northern state of Pará, she cared deeply about the region and the people who lived there. When the public defenders' office of Pará first opened a branch in Altamira in late 2011, she was happy to be reassigned from a smaller city in the region to the dam construction's center of action. As a public defender, Andreia's job was to offer legal support to people in vulnerable situations or who could not afford it, but she went above and beyond. She worked tirelessly to advocate for the people who would be displaced and the farmers, fishers, and others most affected by the dam. She partnered with activists, fellow lawyers, sympathetic government officials, and others who could help support the causes she most cared about.

The high speed at which Andreia was driving confirmed she was tense, slightly nervous that we were running late but more anxious because of the stakes of the "conciliatory hearing" to which we were going. This meeting was intended to resolve demands being made by fishers, boat captains, farmers, riverine communities, and Indigenous people. A month earlier, a small group of two dozen people who fished for a living—none of whom had ever before engaged in political protest—had begun fishing in a prohibited zone near the dam's construction site, as a way to call attention to the changes to the river that were upending their way of life. After three weeks of this protest, other groups joined the fishers. Together, they stormed the construction site under cover of darkness and rainfall, taking control of heavy machinery and ordering all workers to leave. For over a week, they occupied the site, halting construction and calling on high-level officials to hear their demands and provide compensation for the negative effects of dam construction on the lives of traditional communities. In an effort to end the occupation and resume work on Belo Monte, which was Brazil's largest infrastructure project being built at the time, the officials had finally agreed to meet the protesters to discuss these claims.

All the major players would be at the meeting. Top-level representatives from Norte Energia, the construction consortium building the dam, were the targets of the protests. At least one director of Norte Energia would respond directly to the claims protestors were making, in hopes of striking a deal to halt the protest. A federal public prosecutor based in Altamira, along with representatives from Brazil's land reform and Indigenous people's agencies, would facilitate the conversation. Officials from other federal agencies would provide input and expertise. Andreia was charged with representing the non-Indigenous groups, including the fishers, boat captains, and riverine communities. Indigenous leaders would speak for themselves, but they requested that two lawyers affiliated with a nongovernmental organization attend.

I had been to Altamira on four separate occasions since 2010 and had been living there for the four months leading up to the occupation. I rented an apartment in the central part of the city and worked to develop rapport with activists, lawyers, people impacted by the dam, government officials, representatives from the energy consortium, and private companies subcontracted to work on the dam. I developed particularly close relationships with the public defenders and some of the social movements, but I was careful to seek out a broad range of perspectives. I wanted to understand the complexities of the debate over the dam and its impacts, as opposed to the reduced, everyday language of "for or against" the dam that dominated media coverage and surface-level conversations. I attended protests and street rallies, at which I spoke with the leaders and participants, in order to understand the motives, passions, and emotions people had in opposing the dam. I interviewed public defenders and public prosecutors. I contacted prominent business leaders and chatted with small-store owners. I attended the meetings of an innovative forum for sustainable development that the federal government created to bring together regional civil society actors with municipal, state, and federal officials. I spoke with local workers' unions that I knew were important throughout Brazil in supporting marginalized populations. I developed partnerships with professors at the local university who were familiar with the changing dynamics in the region. I joined the fishers as they began their protest and visited them throughout the month, sleeping in hammocks next to theirs on remote islands on the Xingu River.

My efforts to engage a variety of people and perspectives paid dividends. Social movement organizations welcomed me as a member of their groups, inviting me to attend private meetings, cook meals together, and participate in a wide range of other events. Similarly, public defenders gladly allowed me to follow their work, as when Andreia invited me to the hearing and other outreach activities. I also met with a few managing directors of Norte Energia, despite their reluctance to sit down with outsiders, and officials at all levels of government gave me their time. Relationships such as these gave me access to important events like the conciliatory hearing.

In the truck, Andreia turned off the Transamazon Highway onto a recently widened dirt road on which hundreds of small-scale farmers had once lived but which now provided access to many of Belo Monte's construction sites. The roads were in much better condition than when I had visited the area six months previously—a result of the construction process—so Andreia could maintain her high speed, sliding into corners like a rally driver. We soon came to a stop at a checkpoint marking an official entrance to the construction area. One of the guards checked our identifications and explained that we had arrived early. Rather than wait for others to arrive, Andreia wanted to speak with the protestors, in hopes of clarifying the list of demands of those she represented and developing a strategy for the hearing. The guard escorted us through the site, driving his truck in front of ours. We soon reached a long, straight road, which indicated we were on top of the dam that stretched all but a few hundred meters of the nearly four-kilometer width of the Xingu River. To the right of this land bridge, which was upstream of the dam, small islands and patches of forest dotted the expansive river. To the left, the river had dried considerably, leaving undulating terrain but little water. After leading us for nearly two kilometers down this road, the guard stopped to turn around, presumably to avoid conflict, as he told us that the occupiers of the site became agitated when Norte Energia officials or security guards went further.

We continued to the end of the land bridge, where hammocks hung on a few abandoned construction vehicles and two large open-air tents provided shade in the otherwise barren terrain. Andreia handed me a video camera, asking me to record anything that seemed important. As we stepped out of the air-conditioned truck, the sun, heat, and humidity made me feel like I had entered an oven. A few leaders of the group of fishers, a journalist from the south of Brazil, and an activist from a local social movement that had been financially and logistically supporting the protest led us to a small boat that would take us to the nearby fishers' camp on a small island. I had spent two days in the camp the previous week, just after the occupation had begun, and when the boat pulled up to the island a few minutes later, I immediately recognized many of the people on shore. I was struck by the fact that the protest had ballooned from two dozen people to over two hundred and was impressed that they managed to house and feed everybody in their makeshift camp.

Andreia quickly said hello to those she knew, before she gathered twenty people in a circle to finalize their demands, decide who would represent each group, and discuss their strategy. In the list of demands, all the groups included "financial compensation" for loss of work due to the construction. Each population had specific claims as well. Riverine communities wanted official recognition of their lands as traditional communities, which would provide them with additional rights and opportunities, and were asking for compensation for disruption to their livelihoods caused by the dam. People living close to the dam wanted immediate relocation, due to construction-related explosions, increasing security concerns, and transportation challenges. The fishers complained of reduced fish populations, the boat captains wanted compensation for the reduction in their workload, and together, these two groups demanded that Norte Energia stall the final closure of the river until the transportation system was complete and adequate. After Andreia helped the group clarify these demands, they chose about ten leaders to represent the protestors and bring their concerns to the hearing. The meeting concluded, we ate a quick lunch, and we took the short boat ride back to the public defenders' truck.

Back on the land bridge, a minibus had arrived to take the representatives to the meeting. A small group of a dozen Indigenous people and the two lawyers who would provide them legal counsel, as well as a few journalists, had gathered near the minibus. A delegate from the federal police stood with a stern look and a bulletproof vest worn over a dress shirt and slacks. He barked orders in a commanding voice, insisting that only one member of each group could attend. This set off a frenzied debate, with Andreia, the other lawyers, many protesters, and even the journalists pleading that they needed more representation at the meeting. In a symbolic representation of the state's opposition to the interruption of construction and the ability of government actors to dictate the terms on which negotiations would be carried out, the police delegate would not budge on the issue. Each group was forced to leave a number of leaders behind.

The public defenders and I, along with a journalist and the two other lawyers, piled into the truck for the drive back to the other end of the land bridge, where the hearing would be held. One of the lawyers, who worked for a large foreign nongovernmental organization, suggested I remain at the encampment because the Indigenous groups had not authorized me to attend. The public defenders insisted that I stay, once again giving me their video camera and asking me to film the proceedings. I was relieved to have a job to do and grateful the public defenders were helping me.

We drove to the middle of a large dirt area the size of a football field, on which were four open-air tents that would provide shade from the midday sun. The atmosphere at the meeting site was even more tense than it had been at the minibus. There were more debates about who could participate, who would be allowed to attend but not participate, and who was authorized to film the meetings. Andreia insisted that I be able to use their camera, arguing that anybody who wanted to film should be allowed to do so. Norte Energia officials responded that only their camera should be in use during the proceedings. Later, the head of Norte Energia's communication team explained to me that they needed to be careful with who has access to the video of such an event. "It could be manipulated and used in the wrong way," she argued, suggesting they wanted to control the narrative after the agreements were made. The primary Indigenous leader indicated that he wanted restrictions on filming as well, so despite strong arguments from the public defender and journalists, most of us with cameras had to pack them up.

The seemingly small, but intense, debate over who was authorized to record reflected the tone of the proceedings for the subsequent two days. Everyone attempted to exert power at any opportunity. People disagreed about even the smallest of details, and shared as many opinions as there were people present. A few participants who did not know me had suggested that I leave the meeting area, but as I returned the camera to Andreia, she told me to stay and explained to the others that I was accompanying the public defenders' office. After the group decided that I could stay but not film, I went to the back of the tent. Lined up next to me were Norte Energia's director of socio-environmental programs, the police delegate who was still clad in his bulletproof vest, and a woman who looked as stern as the officer. She immediately demanded to know why I was at the meeting, clearly suspicious of my presence. I briefly explained that I was a social scientist from the United States conducting research. In a forceful tone and what seemed like a single breath, she introduced herself as the director of the regional federal police and immigration office and asked me a pointed question: "What type of visa do you have?" Thankful that I had successfully navigated Brazilian bureaucracy, I responded that I had acquired the proper research visa. She quickly said, "Ok. You can stay, but you are not authorized to use a camera and you must remain silent."

It was an auspicious exchange, given that the meeting was central to my research. I could witness direct negotiations between the government, Norte Energia, and groups that felt harmed by dam construction, and the meeting showcased how politics and claims making were being realigned in the region. In the year since construction had started, activists and groups that had worked closely together for decades in struggles for justice and against the dam were no longer speaking with each other. The fishers' protest, the occupation, and its aftereffects would provide clues into the reasons why social and political networks had fractured and how they would reconfigure. It would also provide an opportunity to examine the mechanisms through which new people were engaging and new alliances were forming. Most of the fishers had never been particularly engaged in politics and activism, the public defenders' office was new in the region, and the public prosecutors in Altamira had just begun to play a central role in the conflict. Federal and state government officials had only started paying attention to the region because of Belo Monte but were already deeply enmeshed in the local debates over future development of the region and compensation for the impacts of the dam. I was interested in how these actors that were new to the conflict would engage one another and the longtime activists who had helped shape the region over the previous generation. Additionally, the consortium of public and private entities building Belo Monte was responsible for projects to mitigate the negative effects of construction and compensate people who were adversely impacted. How would this consortium respond to demands? Furthermore, opportunities to deliberate and participate in decision-making forums were becoming ubiquitous, and the use of the rule of law to make demands was on the rise, but how these forms of engagement could lead to change was less understood. The conciliatory hearing provided an opportunity to examine these dynamics and to witness the mechanisms through which people made and responded to claims.

The substantive part of the meeting finally began in the middle of the afternoon, with the representative of FUNAI, the federal agency responsible for policies related to Indigenous people, reading the demands being made by Indigenous groups to everyone in attendance. Each point was discussed in great detail, and as the debates continued, I soon realized that the Indigenous issues would take center stage for the day. The afternoon passed without any discussions of the demands of the other communities, and darkness eventually forced the end of negotiations for the day. After a great deal of debate about where to convene the following morning, everyone agreed to meet at the same location.

A few minutes after we left, Andreia picked up two reporters who had no ride back to the city. Guards then stopped us at a checkpoint to wait for the nightly explosions—used to break up rock—to be completed, as it was not safe to pass any closer. Over the subsequent hour, from 9 P.M. to nearly 10 P.M., a series of loud, earth-shaking blasts made it clear why nearby residents were eager to move. We then resumed our hour-long drive back to the city, but because of the reporters in the car, the public defenders were reticent to share substantive reflections about the proceedings.

The next day, I met Andreia at her office in the late morning, still unsure whether I would be allowed to attend the meeting. Shortly after I arrived, one of her assistants handed me a piece of paper. The one-page document included guidelines for the second day of the hearing, which a judge had outlined, presumably in order to avoid replaying the confusion of the first day. The guidelines indicated that the meeting would focus on the agenda of the riverine and fishing communities, that federal police would once again provide security, that a military police helicopter would be available to provide aerial transport, and, most importantly for me, that outsiders were permitted to attend. Soon after, I was once again in the pickup truck, as the public defenders would not entertain the idea of being transported by helicopter. Andreia was much more relaxed, as was Artur, another public defender. He had learned the first day that he was overdressed with a suit and tie, so he arrived in more comfortable short-sleeved button-down shirt and slacks.

When we arrived at the now familiar blue tents, the atmosphere was also more relaxed than the previous day (though the negotiations would become heated later in the afternoon). Nobody wore bulletproof vests, few requested observers to leave the area, and everybody generally understood how the proceedings would take place. As the official start time approached, people were clearly anxious to begin, but some of the facilitators from the previous day had not yet arrived. I then heard the familiar hum of a helicopter approaching. The last time I had seen this helicopter, it was circling above the protestors just after the occupation started, presumably with Norte Energia officials and police assessing the situation. At that time, a half dozen Indigenous men, their bows drawn, pointed their traditional arrows toward the helicopter as a warning not to land. Now, it landed without incident a hundred yards from the tent, and I expected to see a few federal land reform officials who were assigned to mediate the conversation get out. To my surprise, Thais Santi, the public prosecutor who had only recently been assigned to the agency's office in Altamira and who would help facilitate the meeting, appeared from the helicopter's doors.

In her mid-thirties, Thais came from the south of Brazil. She was filled with energy and determination to support dam-affected people. Like me, her tall frame and light skin made her stand out from most of the people who were from the region. Some public prosecutors were known for their support of marginalized populations, often confronting the state and other powerful actors. Similar to the public defenders' office that Andreia worked in, the public prosecutor's office, while a government-funded institution, operates largely outside of state control. Public prosecutors bring cases against individuals, companies, and the government, often with the goal of enforcing social and environmental regulations and protecting vulnerable groups. Many activists hoped that Thais would become an advocate for the Indigenous and other traditional communities that faced the impacts of dam construction. I had spoken with her on a few occasions, and she seemed determined to align herself with groups fighting for justice and with people facing the worst impacts of dam construction. Given her stance, I was surprised to see her emerge from the helicopter, because many protestors viewed the helicopter as threatening and representative of opposition to marginalized populations. Thais's arrival in the helicopter caused some protesters to worry that she would not become the advocate that people felt they needed. In addition, during the first day of the conciliatory hearing, she had asserted that she was not representing any particular group; rather, she was there to facilitate proceedings, in order to ensure a fair meeting for everybody. She had only recently moved to Altamira, so I wondered if she was unaware of how these behaviors may come across to activists and protestors. Or, perhaps she was skillfully building her credibility and legitimacy, in order to best serve disadvantaged groups in the future. Regardless of her actions in that moment, over the subsequent months and years, fears about Thais's allegiances would be fully allayed, as she became a crucial supporter of dam-affected groups.

Shortly after Thais arrived, the facilitator from the land reform agency announced that the meeting would commence, and they began by addressing each of the six items on the agenda of the traditional communities. The facilitator read the first of the items: "Riverine families living near construction areas should be relocated immediately, due to the explosions, security concerns, and transportation difficulties in those areas." Andreia, the public defender, then prompted Alberto, the unofficial leader of the fishers' protest, to speak on the topic. In his mid-sixties, Alberto had the weathered hands and face of somebody who has spent decades working on the river. Born in a municipality close to Altamira, he moved as a young child to protected Indigenous lands and lived there for over thirty years, as one of his parents was Kayapó. By the time dam construction began, however, he no longer lived on demarcated lands and was not entitled to the compensation Norte Energia provided to Indigenous groups. Instead, he made demands as a fisher, a profession he had been legally registered in for about fifteen years and an activity he had been doing for much of his life. I had traveled up and down the river on his simple narrow wooden boat with a long outboard motor, and it was clear he knew the river like the back of his hand. He navigated the dangerous waterfalls and swirling pools with ease, and seemed content on the water. Alberto was not content, however, with the ways the government and Norte Energia were treating the fishers. According to Alberto and many of the fishers who protested with him, he developed the idea of organizing a group to fish in the prohibited zone and he served as a strong voice throughout the protest, unafraid to confront people. While he spoke and fought for the rights of the fishers, he also embraced his Indigenous identity. At important events like the conciliatory hearing, he wore a beaded necklace and painted lines on the top half of his body and cheeks in a display of his indigeneity.

Alberto got right to the point: "Why haven't you, Norte Energia, compensated these families?" He continued by explaining, in detail, why these problems were so disruptive to everyday life and then noted that the list of affected families was not complete. Andreia added that the fishers demanded that Norte Energia agree to convene more studies and a deadline by which the consortium would establish compensation. Andreia and Alberto used this kind of dialogue to clarify the precise concerns and requests that the fishers were making, as well as to oppose officials when they refused to agree to concrete plans. By working in this way, Andreia was striving to give voice to the fishers, who had been discounted through much of the process.

Later in the meeting, Andreia and Alberto explained the rapidly changing river conditions and argued that Norte Energia should compensate the fishers for lost income. A high-ranking representative from the consortium responded, "We don't have data that says river conditions have changed, and we don't know how the fish population will be impacted." This was a common refrain from Norte Energia officials, who argued that fish populations had been decreasing before construction began and that long-term studies would be needed to determine whether and how much the dam affected fish populations. Alberto quickly rose out of his seat. "That's a lie!" he said, and went on to provide a detailed example of the quantities he used to catch in a week compared to a year after construction began. The Norte Energia official calmly but firmly repeated, "We have *no* indication that things are changing in the river. We have *no* indication of what you are saying, no *evidence* of changes to the fish population in the region." Alberto remained standing to share his frustration with everyone there, exclaiming, "Why does Norte Energia repeatedly diminish the experience of the fishers?"

After a half hour discussing this impasse, it was clear that Norte Energia would not, at that moment, recognize that reductions in the fish population were due to the dam. Thais suggested that the fishers take Norte Energia officials to the specific places where fish numbers had dwindled, explaining that researchers could study those particular areas. Alberto approved of this idea, exclaiming, "Let's set a date right now!" Andreia guided the conversation so that they could determine the number of researchers and fishers that would gather together to carry out the studies. These studies, Thais made clear, would inform future discussions of financial compensation for fishers. Norte Energia was not agreeing to pay any fishers for now.

After the fishers had spent a month protesting, over a week occupying the construction site, and two days engaged in heated meetings, the outcome seemed disappointing but also predictable. It appeared to be another case of a marginalized population courageously confronting a powerful entity to make demands, only for those in power to placate those with grievances by making promises they were unlikely to fulfill. I was thus surprised when, as the facilitators typed up the agreement for both parties to sign, Alberto had a look of relief on his face. As we talked in the fading light of day next to the tent, he continued smiling, and it was apparent he felt he had been victorious. "It's not everything we wanted, but they listened to us," he said. "I think they finally heard us. They were forced to listen to what we are saying." At the time, I could make sense of his celebratory attitude but I did not share his assessment. On the one hand, his positive take was somewhat understandable. Alberto had organized a significant mobilization, brought dam-affected people together, and forced the government and Norte Energia officials to engage in a conversation with them. After a tireless struggle, this moment must have brought great relief. On the other hand, I felt that the fishers had failed to achieve much. Furthermore, it seemed unlikely to me that they would see material or other benefits in the long term.

In fact, time would show that there were many more advancements made during the protest and the conciliatory hearing than I understood in that moment.

INTRODUCTION

After a remarkable three decades of controversy over Brazilian government plans to dam the Xingu River in the Amazon, Norte Energia began constructing Belo Monte in 2011. It became the fourth largest hydroelectric facility in the world when it was completed in 2019. Belo Monte represents a predominant form of contemporary dam construction that is markedly different from what it was in the past. In the middle of the twentieth century, the extent of the negative impacts of dams and other large-scale infrastructure projects were unknown or ignored. Authoritarian regimes constructed many dams, and those in power seldom felt compelled to address environmental and human harm. In contrast, many states in recent decades have adopted a democratic developmental approach, in which they have continued to promote economic growth through large-scale infrastructure projects but have simultaneously aimed to reduce poverty and alleviate the negative effects of these projects on local people.[1] Governments have paired these projects with extensive resources to promote social welfare and introduce participatory modes of governance. This type of dam building undoubtedly represents a step forward in responsible governing. But have such policies really worked?

The years leading up to and during Belo Monte's construction allow us to examine the social, institutional, and political impacts of the democratic developmental approach and how dam-affected populations, engaged citizens, government officials, and other actors respond. The dam, first proposed under Brazil's military dictatorship in the 1970s, came to fruition when the left-of-center Workers' Party pushed it forward after party founder Luiz Inácio Lula da Silva, popularly known as Lula, was elected president. The government saw the dam as a sustainable project that would provide clean energy to the country, and argued that it could be built in a way that would develop the region and minimize the negative effects. As a result, the federal government poured billions of dollars into mitigating its negative socio-environmental impacts; constructing schools, health posts, and other local infrastructure; and integrating more people into decision-making processes with government officials. This book details how a sudden inundation of people, institutions, regional development initiatives, new ideas, and money into a region—along with a major infrastructure project—impacts collective action, relations within and between the state and civil society, and the ability of marginalized populations to fight for social and environmental justice.

For more than a century, outsiders seeking economic and political gain have used and abandoned the region where Belo Monte was built. As a result of the government's failure to attend to the area, local people led grassroots struggles for basic needs, came together to oppose the dam, actively engaged in politics, and organized collectively to improve the social well-being for their communities. All of this, in turn, created strong networks of solidarity. Once dam construction began, just as social cooperation arguably had the highest stakes, these well-established and unified networks fractured. Activists who once worked closely together would no longer speak with one another. Some militants remained fiercely opposed to the dam, while other longtime opponents appeared to accept it, driving a wedge between former allies.

When construction began, the precipitous increase in resources and the new opportunities to participate in governance upended traditional forms of claims making and collective organizing and helped the state and the private sector maintain power and build support for their projects. New investment in the region, coupled with the semi-privatized nature of contemporary infrastructure development, allowed the state to continue to shirk responsibility for the welfare of its citizens. The legal licensing procedures required to build projects such as dams were designed to provide socio-environmental protections, but instead provided justification and support for Belo Monte, regardless of its impacts.

These factors made it more difficult for dam-affected residents and their advocates to voice their concerns to the state through protest and social movements, yet also made way for new and strengthening institutions through which people could make demands. Communities impacted by the dam developed new ways of calling for changes, and despite extensive challenges, they achieved some positive outcomes. Ethnographically examining power struggles in this context and the obstacles that residents and activists had to overcome provides important insights. We see how deep democracy—a set of meaningful, inclusionary, and consequential processes that include but go beyond electoral politics and encourage the state to respond to citizens' claims—can be at once strengthened and weakened in such a context.[2]

We also learn about the changing dynamics of civic and political engagement in Brazil. During the leftist administrations of President Lula and his successor, President Dilma Rousseff, social movements and protest were paradoxically weakened, while processes of participation and negotiation became fundamental in all areas of civic life, and legal institutions took center stage in everything from small-scale fights for rights and environmental protection to the nationwide corruption scandal Lava Jato (Operation Car Wash).[3] The local responses to the changes tell us about the opportunities and constraints facing marginalized populations, social movements, and institutions during and after leftist leadership.

Why should we take an interest in the building of hydroelectric facilities in the Amazon, in how dam-affected populations make demands as their lives change, and in the reconfiguration of politics in Brazil? What do the stories of the fishers, displaced communities, activists, and other residents during Belo Monte's construction tell us that we do not already know about the impacts of state-led infrastructure projects? First, contemporary large-scale infrastructure projects, particularly dams, are different from similar undertakings in the past, so studying Belo Monte sheds light on whether and in what ways the impacts of these new projects are unique. Since the turn of the century, many states have carried out development schemes with purportedly sustainable goals and initiatives that are designed to improve the environmental and social well-being in the projects' regions. The new policies often equate to increased resources and opportunities for local populations. It is thus important to examine how and to what extent the new resources actually mitigate the negative effects. In addition, regardless of the effectiveness of efforts to reduce the harmful trade-offs of development projects, the new policies will affect the social, environmental, and political landscape of a place. We cannot, therefore, examine past projects constructed under authoritarian regimes to understand the consequences of projects constructed under more democratic and politically left-of-center governments. The stories in this book show that today's projects' impacts on local people are different, and often more complex, than similar projects of the past.[4]

Second, a close examination of Belo Monte's consequences, efforts to mitigate its harmful impacts, and the subsequent conflicts provides insights that are particularly relevant in an era of climate change. Not only are hydroelectric dams embroiled in debates over how to reduce the warming of the planet while still powering our energy-intensive lifestyles, but contemporary dams, such as Belo Monte, also offer lessons for strategies designed to *adapt* to climate change. The impacts of dam construction are analogous to effects that a warming planet will have—and arguably has already begun to have—on human populations: forced displacement, deteriorating air and water quality, and public health challenges, among others. Government efforts to manage these effects in the context of dam construction tell us, by extension, about the consequences of state-led efforts to adapt to climate change. As the world experiences more frequent extreme weather events, as sea levels rise, and as local environments are disrupted, many governments, multilateral institutions, and private actors will likely invest in programs to address these challenges and build community resilience. Yet, there is little understanding about the unintended consequences of such investments. This book's account of the Belo Monte project brings our attention to the considerable social, political, and institutional upheaval that accompanies adaptation efforts. The lessons we learn are particularly important in regards to environmental and social justice, as the disproportionate impacts on marginalized communities become clear. This case provides a window into the future of climate change adaptation and helps us better understand the challenges and opportunities that will arise for governments, private companies, civil society, and citizens.

Third, Belo Monte is also worth examining because the conflict surrounding its construction and the subsequent reconfiguration of politics and power struggles give us new insights into contentious politics and the deepening of democratic practices. Scholars have provided excellent analyses of processes of mobilization and resistance on the one hand, and demobilization and accommodation on the other, but often portray them playing out in relative isolation from one another.[5] In the case of Belo Monte, opposition and acceptance, as well as mobilization and demobilization, occurred simultaneously and influenced each other. Furthermore, the multiple avenues through which people make demands show that these dichotomies are not as stark as scholars tend to portray them. Some of the fishers, for example, were neither resisting nor supporting dam construction, but nonetheless organized to make demands. Other residents did not mobilize in the traditional sense, but became involved in participatory decision-making processes. By investigating the complexity of these types of social mobilizations and the ways they affected each other, we can learn why some people engage in civic life while others do not, why some activists change perspectives from resistance to accommodation and others do not, and the impacts of these changes on the effectiveness of claims making.

From this case, we also gain a sense of how contentious politics transform when governments, along with private actors, make new resources available and create opportunities for a wider range of people to participate in decision making processes. Scholars of inequality, social movements, social and environmental justice, and urban change often pay attention to what happens in communities that lack material resources. Social mobilization is usually seen as a process whose goals are to access goods and services, make political gains, and meaningfully participate in democracy. This book shows what happens during and after a sudden flood of resources and opportunities for engagement, as well as what transpires when political allies take office. The results for claims-making efforts, for the broader realm of politics, and for dam-affected communities are not as unequivocally advantageous—or disadvantageous—as scholars might predict.[6]

A close examination of these complex dynamics illuminates how macro-level policy impacts everyday politics and interactions but also how bottom-up practices can reshape institutions. Ethnographic details demonstrate how these back-and-forth processes reconfigured political struggles in Altamira and show the mechanisms through which dam-affected groups, their allies, and others created deeper and more deliberative democratic practices, as well as the barriers they faced in doing so. This account reveals how the work of marginalized groups and their advocates is integral to popular efforts to build a more democratic and just society, particularly as these people engage with the state and other powerful actors. In other words, we see how democracy can be deepened.

STUDYING UPHEAVAL IN EVERYDAY LIFE

The stories that make up this book come from a time of great turmoil and disruption to the lives of thousands of people. In order to understand the range of ways that this tumult impacted the region, I carried out my research with a wide variety of people. Some were well-known activists and government officials, while others were residents who were unknown beyond their own social circles. Some had been involved in the conflict over Belo Monte since the first plans to dam the Xingu were released in the 1980s. Others became involved well after construction began. Some of the people depicted in the book were wealthy and had large social and political networks that were connected to people and institutions in power, while others were humble, had few economic resources, and lacked social and political capital. I interviewed and spent time with anti-dam militants, nongovernmental organizations, representatives from Norte Energia and private companies affiliated with dam construction, government officials, public prosecutors and defenders, and residents who were directly impacted by dam construction, including fishers and people displaced from their homes.[7]

Indigenous groups were key figures in the struggle against the dam. I needed to decide whether and how to include them in my research. After careful thought, I chose to focus on other populations whose stories were not well understood or well publicized. Moreover, an attempt to address Indigenous issues in relation to the dam would require its own study to do justice to the complexities of power, culture, and history within and between Indigenous communities, the state, and broader society. Nevertheless, my research inevitably touched on issues related to Indigenous populations, and some of the people with whom I engaged spoke about and worked alongside Indigenous people.

This range of people and their diverging understandings, perspectives, and experiences are at the center of this research. I was able to more deeply understand the conflict by paying attention to the interactions between these diverse actors and by viewing all of them with the same analytical lens.[8]

In order to capture the voices, lived experiences, and feelings of these people, I spent more than eighteen months in and around Altamira over the course of eight years. I first visited the region a year before construction began, which provided a sense of the place during somewhat typical times and would eventually allow me to see the dramatic changes that accompanied Belo Monte. The majority of my research took place just after dam building commenced, in the middle of 2011, and over the course of the subsequent two years. During that time, I attended rallies and meetings, conducted interviews, and spent time with the people involved in the conflict and other residents. In addition to my time in and around Altamira, I also interviewed government officials, activists, public prosecutors, and others involved and interested in Belo Monte in Belém, the state capital of Pará; Brasília, the country's capital; Rio de Janeiro, a social and educational hub; and São Paulo, Brazil's financial center.

Over the subsequent four years, I followed the processes related to the dam from afar and returned to Altamira in late 2017, more than six years after construction began. While the dam would not be completely finished for two more years, the number of employed workers for the project was a small fraction of what it had been during the peak of construction a few years prior. The population in the area had declined precipitously, and Norte Energia had completed the official resettlement processes for displaced people. Residents in the area referred to Belo Monte's construction in the past tense, indicating that the peak of day-to-day impacts of dam building had ended. The return trip thus afforded me an opportunity to experience life in a new era and learn about the opportunities and challenges that dam-affected populations and others had experienced. In the end, I took nearly 1,000 pages of field notes from over 150 events and interviews.[9] In documenting the stories in the book, I have changed the names of individuals who are not public figures but otherwise have described real people, events, and quotes from my fieldwork.

The use of participant observation, multiple trips to the same place, and extended stays helped me to see beyond common scholarly dichotomies related to large-scale projects in general and Belo Monte specifically. These methods enabled me to examine the complex relations within and between what Stephanie Malin calls "sites of resistance" and "sites of acceptance."[10] Malin uses these ideas to compare how and why communities either reject or consent to nuclear facilities. In contrast, this book's stories about Belo Monte highlight the ways sites of resistance and acceptance are entangled, which influences how demands are made, when mobilization and demobilization occur, and why social and political networks are disrupted. In addition, scholars usually speak of and analyze conflicts over such projects based on the idea that there are two sides: proponents and opponents, or, to use Kathryn Hochstetler's terms, "blocking coalitions" and "enabling coalitions."[11] The two-sided perspective is not inaccurate, per se, but ethnographic work and close analysis show us that the situation is often more multifaceted than "for or against." The reality is that some people move between these two sides, others neither support nor oppose construction but are nevertheless quite active, and yet others focus on supporting impacted communities while remaining ambivalent in regards to the project in general. For many people, once construction begins, the dispute over the project itself becomes irrelevant. Ethnographic research allows us to be attuned to the debate over the enthusiasm for or against the dam, while also aware of the limits to that debate.

My research approach also recognizes that civic engagement, activism, and the everyday practices of democracy involve the interweaving relationships of many varied people. The goal of this work is to examine those relationships, particularly as they manifest *across* the state, civil society, and the private sector. This relational approach to analyzing contentious politics breaks down false scholarly barriers between the state and civil society that tend to obscure the realities of lived experience.[12] As I carried out my research, I paid close attention to the understandings, imaginations, and creativity of people in each situation. Rather than presupposing that people have fixed goals, I drew from the American pragmatist school of thought, which views "human action as a *creative* action" and suggests that people confront problems and are forced to sort out flexible solutions.[13] The attention to the ways people bring their imaginations to bear on the institutions and situations in which they engage allows us to see when, how, and to what extent people are able to influence, and have agency in, their own lives and the broader landscape.[14]

DAMS IN THE ERA OF CLIMATE CHANGE

Climate change is the existential threat of our time. The continued burning of fossil fuels to produce energy, coupled with continually increasing energy demands, will lead to catastrophic and irreversible environmental and social harm. Policymakers, activists, nongovernmental organizations, and everyday citizens are wrestling with how to deal with this crisis. There are two overarching approaches to addressing climate change: mitigation and adaptation. In the broadest and simplest sense, the efforts to mitigate, or limit, the warming of the planet include reducing energy consumption and generating more energy from sources that emit little to no greenhouse gases. The other general strategy in response to climate change is adaptation to the adverse effects that come with global warming. Together, the challenges associated with climate change and the ability, or lack thereof, to deal with those challenges are embedded in an interconnected web of technological innovation; globalized markets; local, national, and transnational political debate; and bottom-up demands for action.

In many ways, dams lie at the nexus of this web, representing the debates about climate change and providing clues into the challenges of addressing the crisis. In terms of climate change mitigation, hydroelectric dams, particularly in the global south, are at the center of the discussion over how to provide the energy required for growing economies while also reducing greenhouse gas emissions. Brazil and other growing economies face increasing energy demands, which are only heightened as millions of people come out of destitute poverty. These developing countries are faced with questions over how to produce more, increasingly efficient, and sustainable energy. For many nations, dams offer a crucial part of the solution. Hydroelectric facilities produce renewable energy without burning fossil fuels, a key factor in reducing global warming. Of course, the social and environmental drawbacks of large dam construction are extensive and well documented, even if some types of hydroelectric facilities, such as run-of-the-river designs, reduce the negative environmental impacts. Additionally, some research suggests that the creation of dam reservoirs produces high concentrations of greenhouse gas emissions from the flooding of vegetation.[15] Given these dynamics, dams are often intensely contested.

The debate is not only about whether or not dams should be constructed, but also includes questions of how adverse effects from construction can be managed. After nearly a century of dam construction that brought vast social and environmental damage, many governments are now building dams with programs and policies designed to address their problems and support local people and ecosystems. In this way, contemporary dam construction closely resembles climate change *adaptation* efforts. The negative impacts of dam construction, after all, are quite similar to the effects of climate change, including forced displacement, deteriorating air and water quality, ecosystem disruption, and public health challenges. The ways dams are constructed and the policies that accompany dam construction can thus tell us a great deal about how to build the resiliency of communities to confront climate change and its effects.

For one, climate change will lead to rising sea levels, increased flooding from heavy storms in some areas, and drought in other areas, all of which will increase migration and likely force millions of people around the world from their homes.[16] The removal of people from their homes and land due to climate change is not unlike the displacement caused by dam construction, the latter of which has forced up to eighty million people globally to move in the past century.[17] Many of the relocation programs associated with dams that were constructed in the middle of the twentieth century under authoritarian regimes were insufficient or nonexistent. The case of Belo Monte provides a window into how the relocation processes might work under democratic conditions. Some scholarship has suggested that such programs can increase livelihood outcomes for residents, while other attempts have led to cultural, financial, and social distress.[18] With the risk that millions around the world will be forced from their homes due to climate change, it is important we assess the direct and indirect impacts of relocation programs. Many of the stories in this book shed light on the resettlement initiatives associated with Belo Monte, the challenges that accompany them, and, importantly, how people are impacted by and make demands around displacement.

Resources dedicated to climate change adaptation programs will, or at least should, go beyond relocation initiatives, as the adverse effects of a warming planet will be far-reaching. The range of adaptation strategies in order to manage air and water quality, protect ecosystems, cope with waste, and maintain public health is likewise vast. The United States Environmental Protection Agency alone has over 150 adaptation actions listed on its website.[19] Scholars, activists, and governmental agencies agree that specific adaptation efforts must be appropriate to a given local context and decided by, or at least in consultation with, local officials and residents.[20] Given the scope of climate change effects and the array of options to address those impacts, the details on *how* to make such decisions, implement programs, and allocate resources are important, yet remain unclear and riddled with challenges. As financial and other types of resources become available, who should be making these important choices? How, if at all, will concerns of equity and justice be integrated into adaptation plans? What impacts will increasing resources have on local communities and their abilities to play a role in decision-making processes?

The case of Belo Monte provides clues. Brazil's federal government invested billions of dollars in programs to manage the impacts of construction and allow, at least in part, local residents and officials to make important decisions for the future. This book shows the difficulties in implementing such programs and lays out some of the unintended consequences of these initiatives.

When we focus on social and environmental justice, as this book does, we further see the important connections to, and lessons for, climate change. In their book *Power in a Warming World*, David Ciplet, J. Timmons Roberts, and Mizan Khan explain that climate injustice, or the "heightened and disproportionate vulnerability to climate-related harm by disadvantaged social groups," occurs not only through the impacts of climate change but also in the efforts to respond and adapt to the adverse effects of a warming planet.[21] In the same way, both the harmful impacts of large-scale infrastructure development *and* the attempts to lessen the adverse effects of those projects disproportionately affect marginalized communities. When dams are built, the fate of people with little economic, social, and political capital tends to be determined by whatever plans the government or other actors have, or have not, designed. With few resources, disadvantaged communities can struggle to cope with displacement, rising costs of living, increased rates of violence, and a strained infrastructure, among other effects of dam construction.[22]

The stories in this book bear out these challenges and also highlight hardships that can accompany efforts to manage the undesirable consequences of environmental change. We find that despite, and in some ways because of, the increased resources dedicated to addressing negative impacts of Belo Monte's construction, underprivileged groups faced significant distributive, procedural, and recognition-based injustices. Historically, analyses of environmental justice efforts focused on distributive issues, which center on who suffers burdens related to environmental change and on other material outcomes, such as who receives compensatory benefits. More focus has since been placed on procedural justice, which refers to how decisions are made and by whom, and recognition-based justice, which hones in on particular identities, needs, and capabilities of specific groups. Scholars have clearly articulated why all three of these notions of justice should be fundamental in our analyses of the impacts of environmental change caused by climate change and other disruptions, as well as adaptation strategies.[23]

This book makes clear that significant financial and institutional resources—as well as state-created mechanisms to engage more people in decision-making processes—are necessary, but not sufficient, to address injustices against marginalized communities. In the case of Belo Monte, the state provided dam-affected populations with avenues to work toward all three forms of justice, yet these opportunities did not ensure positive outcomes on any front. Instead, possibilities for these forms of justice to emerge only came about with a combination of top-down and bottom-up work. These struggles remind us that increases in resources dedicated to climate change adaptation are, and will continue to be, managed through local structures of power relations between government officials, impacted communities, other engaged residents, activists, and transnational actors. These struggles for power reconfigure social and political networks and the opportunities that people have for making claims. National and international strategies to adapt to climate change will impact local outcomes, but local processes will also impact the possibilities of achieving distributive, procedural, and recognition-based environmental justice on national and global scales.

DEEPENING DEMOCRACY AND RECONFIGURING POWER STRUGGLES

By paying close attention to the everyday politics and practices in the context of the upheaval caused by Belo Monte, this book addresses key questions about power and democracy. How do subordinate groups challenge power structures? How can citizens participate in ways that matter for the communities in which they live? When the left comes to power, as in Brazil, how and why do leftist forms of activism change?

Revolutionary thinkers, much of the social movement literature, and activists on the left have long celebrated protest as a primary path to challenging structures of power. Many believe that everything from small-scale reforms to the toppling of regimes is possible when enough people with clear demands "go to the streets." These actions could bring about radical visions of democracy and give way to new political structures. But given the challenges that activists have in securing even the smallest of victories, let alone revolution, some scholars and activists are increasingly skeptical of the potential of traditional protest activities in modern political regimes.[24] Their arguments resonate with those developed by Antonio Gramsci nearly a century ago. Gramsci argued that "frontal attacks" are futile and contended that oppositional and revolutionary actors must instead engage in a "war of position," working to create new power structures in the terrain of culture and ideology.[25] In these conceptions, claims making based on direct action is largely ineffective.

The stories in this book suggest that these analyses are partially true. The long history of the conflict over damming the Xingu, which began in the 1970s, provides insights into how and why the impacts of protest have changed. In 1989, abundant local, national, and global protest activity contributed to halting initial plans to dam the river. As revised plans progressed in the 2000s and as construction appeared more certain, traditional activism weakened. Protests failed to accomplish the demands of activists, and radical notions of democracy and reimagined visions of society were often lost. Whereas the authoritarian regime could be threatened by large-scale and widespread protest activity, the left-of-center democratic government was, paradoxically, largely insulated from the political effects of direct action.

Protest was also undermined by broader structural forces. The global flows of ideas, money, power, and imaginaries, in which Belo Monte was tied up, also made protest ineffective. The drive for, and rhetoric around, sustainable development, economic growth, and national energy independence were, and continue to be, invoked in creating and maintaining the global hegemonic order. Hegemony, for Gramsci, develops as ruling classes successfully establish connections between their shared interests and ensure consent from subordinate classes through cultural, moral, and ideological leadership.[26] In this case, the powerful global actors that had interests in building large-scale infrastructure projects created enough consent, and had enough material resources, to make rallies and marches futile in their attempts to stop construction. Belo Monte was in fact built, a sign for some scholars that activists failed and their methods were ineffective.[27]

This book's account of power struggles, however, does not stop there. Even as traditional protest was largely ineffective on its own, the conflict around Belo Monte suggests that protest still plays important, albeit constrained, roles in the struggles of marginalized communities. The stories about this conflict illuminate the ways protest can shape if and how people gain access to other ways of making change. The stories also show how the increase in opportunities to participate in decision-making forums impacted and was impacted by protest, collective organizing, and other avenues. The fishers, for example, were able to use a single, protracted public demonstration to jumpstart their opportunities to participate in decision-making processes. They were able to construct social networks that gave them access to government and legal institutions, which could help them negotiate for better outcomes. This and other examples of direct action did not threaten state power or government projects, as was the case in the past, but public demonstrations did shape state-created participatory forums and the use of the law.

Scholars, activists, and government officials, particularly those on the left, have long been interested in whether and how democracy can go beyond electoral politics. Classic scholars and philosophers of democracy, such as Alexis de Tocqueville and John Stuart Mill, argued that associational spaces and extensive participation in politics and decision-making are important for the overall health of democracy.[28] In the last few decades, this interest has developed into an expanding practice of, and research in, deepening democracy, which John Gaventa defines as "a school of thinking that focuses on the political project of developing and sustaining more substantive and empowered citizen participation in the democratic process than is often found in representative democracy alone."[29] After democratization swept through Latin America in the 1980s, left-of-center governments began instituting experiments in participatory decision-making processes.[30] Brazil's participatory budgeting initiative, which gives citizens control of a portion of municipal budgets, became one of the first of the new strategies to incorporate more people into direct deliberation with fellow citizens and government officials. Participatory processes have since expanded around the world and into the health and education sectors, the management of natural resources, and community development efforts.

Much of the literature on deepening democracy has examined these institutionalized examples of "empowered participatory governance," and a significant debate over the value of these spaces has emerged.[31] Similar to the skepticism toward protest, some scholars, activists, and policymakers who once heralded participatory decision-making efforts as an important way to distribute more power to the public increasingly see these opportunities as fraught with problems. These efforts can co-opt radical visions of democracy, divide previously unified subordinate classes, serve as a justification for states and powerful economic actors to carry out their own agendas, and lead to the "democratization of inequalities."[32] Despite the criticisms, participation is still celebrated by many and remains central in development planning. In fact, Belo Monte is one of the first large-scale infrastructure projects to include well-funded and well-attended participatory forums of negotiation. While some scholars and activists have critiqued—and even disregarded—these opportunities as another way in which the state has bought off local communities in order to gain consent, this book shows they are consequential and not to be dismissed.

Not only are the processes within these participatory spaces telling of local democratic practices, but we also gain important insights by investigating how official participatory opportunities interact with other avenues of engagement, including protest, the law, legal institutions, informal spaces of negotiation, and other forms of collective organizing. Empowered participatory governance efforts are not, of course, the only forms of participation that can encourage and lead to deeper democratic practices. Yet, while there is a great deal of literature that examines other ways of making claims, such as through direct action or the legal system, scholars seldom analyze the interactions between various types of participation and how these interactions impact the constraints and possibilities for more inclusive democratic practices.[33] By studying the full scope of engagement, we better understand that power struggles—as well as social and political relations—are reshaped as the avenues through which people negotiate, participate, and make demands influence and are influenced by each other. This approach also encourages us to step back from assessing the impacts of participatory governance and examine how democratic practices come to be. The case shows how local democracy is co-constructed by multiple actors from both the state and civil society and through multiple forms of engagement.[34]

The comprehensive examination of engagement efforts also allows us to see the extent to which the logics of participation, deliberation, and negotiation permeated into all spheres of civic and political life. This resulted in what I call hyperdeliberation, or the ubiquitous, necessary, and often excessive processes of deliberation to make and settle demands. On the one hand, participatory processes became meaningful avenues of engagement for marginalized communities and on occasion even reoriented power dynamics. On the other hand, deliberation and negotiation were often necessary burdens, rather than suggestive of a more procedurally just approach to development and decision-making. Broad acceptance of deliberation strained dam-affected people with additional demands on their time and resources and forced them to rely on mediators to advocate for their needs. Additionally, many residents lacked the experience and knowledge to be equal participants with government officials and Norte Energia representatives.

The efforts toward social justice in this context are further complicated as deliberative processes rely on translators and mediators of claims, something we clearly see in the case of Belo Monte. I use the terms "mediation" and "translation" to highlight processes that are central to practices of local democracy. Drawing on Ann Mische's research on Brazilian youth activists, I define mediation as a communicative practice in which an individual or an organization transmits demands between other disconnected groups, a process that "involves negotiating between multiple possible public representations of who one is acting 'as' as well as what one is acting 'for.'"[35] Translation can be understood as the process whereby actors partially external to a group or a space of negotiation reframe people's words and needs into demands that can be acted upon. We can think of translation as a type or form of mediation. Scholars have argued that the mechanisms through which mediation and translation occur are important causal determinants of whether and how local democratic deepening occurs. As Gianpaolo Baiocchi, Patrick Heller, and Marcelo Silva point out in their book about participatory budgeting in Brazil, mediation can take forms that degrade democracy by increasing inequalities, such as clientelism, as well as forms that enhance democracy, such as those that promote cooperation between associations and prevent any single individual from maintaining control.[36] In this book, I pay most attention to the mediators and translators of claims that have the potential to enhance those claims, while also acknowledging the precarity of groups that must rely on mediation.

In the case of Belo Monte, officials from legal institutions, particularly the public prosecutors' office and public defenders' offices, emerged as the primary and most effective translators and mediators of claims. They spoke with and for communities, such as the fishers, who were impacted by dam construction but who were unable to make themselves heard through other means. Despite occupying prominent positions within the legal system, public lawyers only made significant gains for marginalized communities outside of the courts. This particular use of legal institutions provides an important insight into our understandings of the ways social movements and struggles for justice use the law outside of litigation, an unexplored topic in the literature. In addition, this form of legal mobilization shows an important reconfiguration of politics and reshaping of power struggles, highlighting surprising possibilities for marginalized communities to make demands and receive at least some compensation. At the same time, this case makes clear how ineffective the legal system can be and the tenuous situation that many communities thus encounter in the face of social and environmental upheaval.

As the stories in the book will show, progress by marginalized groups in seeking justice relied on slow processes that involved a series of actions by and between activists, researchers, community members, and legal officials, all of whom also negotiated with government officials and private actors at every step. The details of these processes highlight that, while the goals of struggles for justice and rights may often be material, the terrain of struggle is over recognition, knowledge, legitimacy, and authority. This terrain is continually changing, particularly as new, substantial resources and opportunities emerge. Through a close examination of these changes, we can see how people's struggles might produce a more inclusive and socially just democracy.

PART I   HYDROPOWER, RESISTANCE, AND THE STATE

1 · DAMS AND DEVELOPMENT

Less than two months after construction began, elected officials from the municipal, state, and federal levels held an event in Altamira to commemorate the groundbreaking of Belo Monte and the social programs and mitigation efforts that accompanied it. It was early in my fieldwork, but I had established connections with local officials, social movement leaders, and freelance journalists. I found myself attending the event's press conference as the photographer for a journalist. To reach the entrance, we had to make our way through a throng of anti-dam protestors who shouted, banged on drums, and chanted slogans such as "Fora Norte Energia! Queremos Moradia!" ("Go away, Norte Energia! We want housing!"). When we reached the entrance, the guard showed us the way into a two-room building and said the event was about to begin. We had missed the social hour, evidenced by the nearly empty platters and glasses on a long table we passed by as we walked through one small room and into another. There, approximately thirty journalists from local, regional, and national press outlets crowded together, waiting with their cameras and notebooks. Four men dressed in suits with earpieces looked to be the security detail for the government officials, and a handful of armed police officers were scattered around the room. About six microphones and a few small audio recording devices were on the table at the front of the room, behind which sat the mayor of Altamira, two federal ministers, the state minister of energy, and a high-ranking representative of Norte Energia. I managed to place my microphone on the table just as the press conference began.

While the protests continued outside, the government officials inside touted the benefits of the project for the people living in the region. Zé Geraldo, a federal congressman from the Workers' Party who had previously been an activist along the Transamazon, stressed the government's commitment to the people: "More important than the installation of the dam are the social protection and promotion of residents [in this region]. Belo Monte is important for Brazil, important for [the state of] Pará, but our people are much more important." He concluded with a comment on how this dam project could be different than those that had come before it: "Belo Monte serves as an example to the world that yes, it is possible to construct a project of this magnitude in the Amazon while respecting the citizens, respecting the environment, and respecting all of the Brazilians who have come here."

The contradicting sentiments inside and outside the press conference highlighted the conflict surrounding Belo Monte, which, in many ways, was nothing new. For decades, dams have been at the center of debates over the best ways to generate development. On the one hand, dams have come to symbolize progress and modernization, as they epitomize advanced engineering and the ability of humans to control nature. On the other hand, the social and environmental impacts that accompany dam construction have become representative of the problems that come with unbridled drives for development. Furthermore, the kind of development that the state pursues through large-scale infrastructure projects is often at odds with the development that activists call for, such as the safe, secure, and stable housing that the protestors were referencing in their chants. Like many other places in the world, dams have been integral to the push for development and imaginaries of progress in Brazil. In his comments, Congressman Geraldo was alluding to, and recognizing, the destructive legacy of dams and the competing notions of development. He was arguing that Belo Monte could be constructed in new ways that supported the population and that would show the world that large-scale projects could achieve broad development objectives, such as economic growth and energy independence, without the downsides that have plagued such projects for generations.

Most development projects come with trade-offs of some kind, and therefore produce conflict. Energy generation always has ill side effects, industrial production brings pollution and waste, and economic growth is often accompanied by increased inequality. Given dams' purported benefits to local and national economies and their symbolism of progress, along with their negative consequences and the resistance that surrounds their construction, dams epitomize the challenges of and the controversies surrounding development. As the World Commission on Dams summarized, "The debate about dams is a debate about the very meaning, purpose, and pathways for achieving development."[1] Dams are thus ideal projects through which to examine the trade-offs of development, the ability of states to mitigate undesirable outcomes, and the potential for those negatively impacted by development projects to have a voice in the process and lessen the adverse effects.

CONTRADICTIONS AND CONFLICT

For millennia, dams have been used to control the flow of rivers and lakes, prevent floods, and provide dependable sources of irrigation for agricultural crops. Hydroelectric technology emerged in the late 1800s, and large dam construction grew throughout the twentieth century, first in the United States, Canada, and

Europe, and then in countries throughout the rest of the world. For proponents, dams, especially hydroelectric dams, are the holy grail of development. Not only do they generate necessary and purportedly clean energy to power modern, industrial economies, but they also provide a number of other benefits. They can create large numbers of jobs during construction, temporarily boosting regional economies. They can also generate income at the national level through opportunities to export electricity and through the goods produced by electricity-intensive industries, such as aluminum refining. As nations seek energy independence and self-sustainability, hydroelectric dams can play a significant part in reaching those goals. Furthermore, proponents of dams argue that the energy that hydroelectric facilities produce is renewable and sustainable, thus helping us move away from reliance on fossil fuels and toward a greener economy that can help mitigate climate change.

National leaders, the energy sector, and developers have used the rhetoric of development and progress to justify dam construction since the beginning of the twentieth century. In 1932, U.S. president Herbert Hoover, when visiting the construction site of the dam that would ultimately bear his name, captured the sentiment of proponents when he remarked on the dam's significance: "The waters of this great river, instead of being wasted in the sea, will now be brought into use by man. Civilization advances with the practical application of knowledge in such structures as the one being built here in the pathway of one of the great rivers of the continent. The spread of its values in human happiness is beyond computation."[2] Hoover's successor, President Franklin D. Roosevelt, likewise said dam construction provided opportunities to use otherwise wasted water resources for irrigation, flood control, and hydroelectricity generation. He also stressed that dam construction would create jobs for many people left unemployed by the Great Depression. These arguments remained dominant over the subsequent three to four decades, and the number of dam projects increased in the United States, Canada, and Europe until the 1970s.

Other leaders of large nations similarly saw dams as an important tool in modernization efforts in the early to middle part of the twentieth century. The first provisional president of the Republic of China, Sun Yat-sen, imagined damming the Yangtze River in the early 1900s to generate electricity and protect communities from floods.[3] Nearly a century later, after China had already built more large dams than any other country, Sun's ideas came to fruition when the Three Gorges dam was completed in 2012, becoming the world's largest dam (and displacing over a million people, far more than any other dam).[4] Large dams have also been central to development policy in India. In 1954, the prime minister, Jawaharlal Nehru, famously celebrated the damming of rivers by calling dams the "temples" of modern India.[5] By 2018, nearly five thousand large dams had been built in the country, including in parts of the Narmada River complex, one of the largest dam projects in the world, which will consist of thirty dams if completed.[6]

The perspective that dams symbolize progress and produce clean energy has had tremendous impacts on the scale of dam construction. More than fifty-nine thousand large dams have been built worldwide, across more than one hundred countries, and hydroelectric facilities have become a central component of the energy matrix in a wide array of nations. A third of the world's countries generate over half of their electric power from dams, and at least a dozen countries rely almost exclusively on dams for their electricity. Hydropower makes up 76 percent of all renewable electricity in the world. The extent of dams is noteworthy on its own, but a distinct geographic pattern has emerged in the last four decades. In North America and Europe, the dwindling of suitable sites and the rising concerns over social and environmental impacts led to a decrease in dam construction—and an increase in dam removal—at the end of the twentieth century and into the twenty-first century.[7] In contrast, dams have remained central to development objectives throughout the global south, with construction rapidly increasing in those regions since 2005. In addition to the large countries that are seen as the leaders in hydropower, such as Brazil, China, and India, nations as diverse as Argentina, Pakistan, Turkey, Mozambique, and Egypt, among many others, have continued building large dams.[8]

The powerful representation of dams as sustainable progress has been countered by environmentalists and dam-affected populations around the world for decades. Opponents cite the negative environmental consequences, the unequal distribution of the benefits and burdens, many countries' authoritarian approach to construction, and the endemic corruption that often accompanies dam planning and building. Ecologically, dams disrupt rivers in a number of ways. They divide a single ecosystem into areas above and below the dam, isolating species and blocking migration patterns. Early and mid-twentieth-century dams on the Columbia and Snake rivers in the United States, for example, destroyed salmon and steelhead populations by blocking their spawning runs. Dams also cause rising and falling sediment loads along the banks, riverbeds, deltas, and estuaries; significantly altered water quality, including flow, nutrient load, and temperature; and the elimination of natural flooding, which some scientists suggest is the greatest harm dams pose to watersheds. While some species may benefit from these changes, dams lead to overall decline in biodiversity and a harmful fragmentation of ecosystems. Furthermore, the flooding of previously dryland vegetation creates high levels of greenhouse gas emissions, contradicting the arguments that dams can help mitigate climate change.[9]

Dam construction has also led to adverse social and economic impacts around the world. The most discussed, and often most disruptive, impact of dam construction has been displacement. Dams' construction sites, river diversions, and the flooding of lands to create reservoirs have led to the removal of an estimated forty to eighty million people from their homes around the world. Most individual dam projects force the relocation of tens of thousands of people, while about a dozen have displaced more than one hundred thousand each.[10]

While forced displacement is always a serious consequence and one of the first concerns related to any dam project, there are numerous other social impacts of construction. Rapid population growth, due to increased yet temporary economic opportunities, strains local infrastructure and raises crime rates and the cost of living. Prostitution rings, drug trafficking, increased violence, and abuse against women also tend to accompany dam construction. Despite all the ways communities closest to hydroelectric dam construction are negatively impacted, the electricity is often distributed elsewhere. The people and the microclimates closest to the dams disproportionately bear the burdens of dam construction and often do not reap the rewards.[11]

These negative impacts have spurred collective mobilization efforts against dam construction around the world. For decades, the people impacted and potentially impacted by dams have come together to demand better compensation, and coalitions of supporters have sprung up in many countries to pressure their governments to halt building. International groups opposing dams have also joined grassroots anti-dam movements, becoming major players in the conflicts over dams. Professionalized nongovernmental environmental organizations, such as International Rivers, Amazon Watch, and Greenpeace, have pressured governments and companies to halt these projects, supported local movements, advocated for dam-affected people, and influenced relevant legislation. The Narmada Valley project in India, for example, faced local and transnational resistance for decades. Likewise, protest against dams in Latin America and Africa have become symbols of campaigns aimed at protecting the world's rainforests.[12] Even when these coalitions fail to block construction, they often slow down the process, giving time for mitigation efforts to take effect and leading to better outcomes for the people affected.[13]

DAMMING BRAZIL

This history of dams in Brazil reflects global patterns of dam construction, including these same conflicts and controversies. Dams, particularly hydroelectric dams, have been central to Brazil's developmental ambitions for decades. Extensive dam construction over the past century has made the country the world's second largest producer of hydroelectricity behind China. Prior to Belo Monte's construction, hydropower already made up nearly 30 percent of the country's energy supply and more than 80 percent of its electricity. Including Belo Monte, Brazil is home to three of the world's seven largest dams.[14]

Brazil's rise as one of the world's leaders in hydropower is rooted in the country's long-standing, broader view that industrial growth and infrastructure projects are key to development. These ideas date back to President Getúlio Vargas, who led Brazil for a total of eighteen years between 1930 and 1954. Both his authoritarian and democratically elected administrations pushed industrial advancement, encouraging industrialization in the south and southeast and spurring the construction of dams throughout the country.

Through these activities, the foundations of a developmentalist paradigm in Brazil were forged. A developmental state is one in which the government significantly intervenes in the market to spur rapid industrial and infrastructure growth. In developmental states, the social and environmental effects of large-scale projects are largely ignored, as elite bureaucrats make many of the economic decisions without the input of civil society actors or even many political leaders. In other words, developmental states tend to advance their agendas in authoritarian regimes in which civil society is relatively weak, either because it was weak to begin with or as a result of suppression and disregard of public opinion. This configuration has allowed central governments to construct large dams, among other projects, without concern for the social and environmental impacts of those projects.[15]

Shortly after taking control of the Brazilian government through a coup in 1964, the military regime consolidated its developmentalist approach. The image of dams as symbols of national independence and modernization aligned with the dictatorship's nationalist goals, as hydroelectric power could help satisfy growing energy demands for the rapidly industrializing country. Some of the first hydroelectric dams had already been built in the south and southeast of Brazil from the 1930s to the 1950s, and the sizable Furnas Dam on the Rio Grande was initiated in 1958, but most of the largest dams were constructed, or at least designed, under military rule between the 1960s and 1980s.[16]

One of the most significant of these large dams was Itaipú, a dam on the border of Paraguay and Brazil. With a generating capacity of fourteen gigawatts, Itaipú became the largest hydroelectric facility in the world when it was completed in 1984, less than ten years after construction began. The dam had significant social and environmental costs. It flooded over a million square kilometers of land. Not only did this flooding submerge some of the world's most beautiful scenery, but it also displaced over forty thousand people and forever altered the regional ecosystem. In addition, at the peak of construction, nearly forty thousand workers were employed to build Itaipú. Most of these workers migrated from other places in Brazil and Paraguay, which led to a host of other problems. The population of the small towns near the dam site increased up to sevenfold, leaving the workers and their families to live in crowded conditions with substandard sanitation and going without adequate health care. The rates of infectious disease, malnutrition, and sexually transmitted diseases were high, and malaria rates increased dramatically in the region, which had seen near eradication of the disease prior to construction. The scope of the project—and perhaps the scant attention the government paid to the negative impacts of the dam—make Itaipú one of the most symbolic projects of the military regime's fraught developmental approach.[17]

Hydroelectric facilities also aligned with and supported the military government's goals of frontier expansion into Amazonia. Dams on rivers and tributaries in the Amazon could power new energy-intensive industries in the region. These industries relied on the extraction of raw materials that were widely available in Amazonia. The new transportation routes made available by the Transamazon Highway and other roads, new shipping ports, and airports allowed for the movement of goods and people necessary for the growth of industry. These congruent benefits of regional growth and the concentration of capital intensified state-led industrialization policies. This approach to development further centralized the planning process, benefitting the interests of elite networks, business interests, and industry. Given these dynamics, the planning of new dams on the rivers and tributaries of Amazonia began early in the dictatorship, and construction of megadams in the region commenced by the mid-1970s.[18]

In 1975, Eletronorte, a regional subsidiary of Brazil's state-owned energy company Eletrobrás, began building one of the largest of these dams, Tucuruí. Located on the Tocantins River, Tucuruí is just over two hundred miles east of Altamira. The first phase of construction, which included the main dam and led to the majority of impacts, was completed in 1984. Tucuruí provided energy for industrial projects, which the government ensured by providing subsidies to foreign aluminum smelting companies. As a result, approximately 65 percent of the dam's energy from the first stage of construction went to two foreign-owned aluminum firms in the states of Pará and Maranhão. By 1991, these smelting plants were consuming 3 to 5 percent of Brazil's total electrical energy. As Philip Fearnside shows, the subsidies tied to Tucuruí's power supply distorted the entire Brazilian economy, by providing nearly free energy to foreign companies that were using a significant share of the country's available energy. The use of energy by these aluminum plants also allowed the government to argue that more dams needed to be built in order to power the cities of Brazil.[19]

Tucuruí's first stage of construction brought a host of negative consequences to the region's social and environmental landscape. The number of displaced people, 25,000–35,000, was substantially higher than the original estimates, and 3,700 people were displaced twice, because Tucuruí's reservoir flooded their original resettlement locations. In addition to problems with resettlement programs, construction caused serious health impacts that were not considered risks during the planning process, including a rise in malaria and other mosquito-borne illnesses, accidents, and sexually transmitted diseases. With the drastic change in the flow of the river, local ecosystems were fundamentally altered. This decimated fish populations and forced some people who relied on fishing for income and sustenance to change professions or leave the region. In fact, some moved west to the Xingu River, only to face the threat of another dam. Research also suggests that the flooding of the vegetation to create Tucuruí's reservoir caused more greenhouse gas emissions than would have come from thermoelectric power plants generating the same amount of electricity as Tucuruí. These emissions were also more than the total fossil fuel emissions of the city of São Paulo. These emissions were likely heightened because Eletronorte chose to clear only 30 percent of the vegetation, despite recommendations that 85 percent should have been removed prior to flooding. In order to highlight this flooding of vegetation and the waste of resources, artists and journalists captured striking photographs of Tucuruí's reservoir filled with upright yet leafless trees, and of men using chainsaws underwater to cut floating trees and logs.[20]

These impacts resulted from the dictatorship's lack of detailed planning, its disregard for the regional concerns and needs of the people directly impacted by construction, and no concentrated effort or political will to develop the local infrastructure. The government's close connections and desire to develop industries such as aluminum and metallurgy meant that the business leaders running these industries, as well as the major construction contractor, Camargo Corrêa, were some of the only voices heard during the planning stages. Few other groups were able to contribute to decision-making processes, there was virtually no consideration of local knowledge or ideas for how an abundance of new resources to the region could best be used to support local communities, and no mechanisms existed through which citizens could participate in decision-making. Even people who were directly affected by the construction had little participation or even understanding of the situation. The government and construction companies gave hardly any information to the affected populations, and the authoritarian ways the dam was planned meant that many groups were, as Henri Acselrad summarized, "politically disqualified and culturally deprived of their character."[21] It was only after massive protests in 1981, six years after construction began, that Eletronorte created a committee to consider thousands of complaints from dam-affected people, and many of these cases went unresolved. Furthermore, despite the increased energy to power aluminum smelting plants, many towns, villages, and rural residents close to Tucuruí—some within sight of the electricity transmission lines—went without power for more than a decade. In 1997, and only after public outcry, Eletronorte finally met with local town councils to sign an agreement to construct a step-down substation for the town of Tucuruí and other communities, which had been receiving unreliable energy from a branch line on an Eletronorte work yard.[22]

Tucuruí was poorly planned and led to a host of negative consequences, but the Balbina dam in the state of Amazonas was arguably worse. Not only were the impacts of construction socially and environmentally destructive, but the dam also produced very little energy. During the late 1970s, the government claimed that Balbina would supply energy to the growing city of Manaus at a time when oil prices were high and transmitting energy across long distances was quite difficult. The dam could also facilitate mineral extraction nearby and potentially bring jobs and resources to the region. Whatever the motivating factors, the military dictatorship pushed this type of project forward—without much resistance—as part of its larger image as a modernizing and industrializing country. However, as in the case of Tucuruí, little forethought went into Balbina's plans, and very few people were incorporated into decision-making processes. Construction started in 1981, two years after originally planned, but little to no research regarding the technological feasibility and effectiveness of the dam had been done, even though scientists knew there were significant flaws in its plans. While the reservoirs of Balbina and Tucuruí are approximately the same size, Balbina's installed energy-generating capacity is about 3 percent that of Tucuruí.[23]

Balbina faced significant criticism from the early stages of planning because of the disproportionate amount of flooding and other effects it would have for the relatively little amount of energy it would supply. In the end, Indigenous and traditional communities, as well as peasant farmers, were forced off their lands. Some Indigenous groups became entirely dependent on the government for basic provisions, peasant communities were relocated to lands that were isolated and difficult to farm, and many of the displaced were not given secure land tenure, making it less likely that they would make long-term investments in their land. Environmentally, swaths of forests were removed, ecosystems were disrupted, some species were threatened, and, as with Tucuruí, the flooding of the reservoir led to soaring carbon emissions. By most accounts, Balbina was a disaster. Even some proponents of hydropower in the Amazon, who have staunchly defended Tucuruí as showcasing exemplary development, have since suggested that Balbina was problematic and should not have been built.[24]

During most of the dictatorship, the authoritarian government effectively prevented organized resistance to dam construction and made few concessions to affected populations. The first social and political openings for a stronger social movement against dams emerged in the late 1970s. Authoritarian rule had begun to relax through political programs known as *abertura* (opening) and *distensão* (decompression), and pressures on the military government to return the country to democracy increased.[25] Furthermore, the negative impacts of large hydroelectric projects became clearer, and dams faced increasing criticism among environmental activists, impacted residents, and engaged citizens from around the world. In 1979, when the government released plans to build over twenty dams in the south of the Brazil, progressive parts of the Catholic Church, rural unions, colleges, and activists came together to demand that people affected by the dams be fairly compensated for the ways their lives would be uprooted. Soon after, the groups began to oppose the dams outright, and they officially formed the Comissão Regional de Atingidos por Barragens (CRAB—Regional Commission of People Affected by Dams), marking the birth of the anti-dam movement in Brazil. CRAB and other regional anti-dam groups that emerged around the country in the early 1980s would eventually come to play crucial roles in the creation of the nationwide organization Movimento dos Atingidos por Barragens (MAB—Movement of People Affected by Dams) in 1991. MAB would become the most vocal and well-known group that resists dam construction and fights for the rights of those affected by dams around the country. The movement strengthened by building strong bases of support at the local level and developing connections to other national and regional social movements, international activists, and the Brazilian government, particularly the Workers' Party.[26]

At the same time that the national coalition of dam opposition strengthened, Brazil's efforts to finance large infrastructure projects became more difficult. The debt crisis of the 1980s in Brazil slashed available spending for such projects, and the World Bank began reducing its support for large dams, in part due to pressure from outside groups that highlighted the social and environmental harms. As a result, dam construction slowed from its peak in the 1960s and 1970s, when Brazil built approximately one hundred dams in each decade. By the 1980s, the number of new dams was down to about sixty, and fewer than thirty were built in the 1990s.

DAMMING THE XINGU

The history of large hydroelectric dams in Brazil and around the world shows the destructive power that such projects can hold and the conflicts that ensue. Given the impacts, it should be of little surprise that many such dams were constructed under authoritarian regimes and that dam construction in Brazil and elsewhere waned in the 1980s and 1990s, when the country democratized again. Instead of addressing the negative social and environmental impacts of these projects, developmental states, like that of Brazil under military control, maintained strong central governing power that restricted civil society participation, subdued public outcry, and lessened their need to respond to the demands made by dam-affected citizens. This allowed governments to enact their development agenda with little resistance. It is thus significant that the rates of dam construction in the world began to increase again in the twenty-first century and within many democratic countries with left-of-center governments. The pressing questions for governments, dam-affected residents, and scholars who study development and dams in this context are about why this uptick occurred and how to manage the negative impacts that accompany dam construction.

The Belo Monte hydroelectric project is a particularly apt case through which to examine these questions. The remarkable decades-long story of damming the Xingu River shows how the Brazilian governments' approaches to and conflicts surrounding dam construction have changed. In the 1970s, the authoritarian military dictatorship introduced the idea of building six hydroelectric dams on the

Xingu and connected Iriri River. Throughout the decade, the Consórcio Nacional de Engenheiros Consultores (CNEC—National Consortium of Engineering Consultants), which would continue to play a central role in planning and implementing Belo Monte decades later, carried out a study to determine the viability of damming the Xingu. CNEC completed the first outcomes of that study in 1979, although plans were not released to the public until the end of the 1980s. The plans called for a complex of six dams—Kararão I and II, Iriri, Jarine, Kakraimoro, Ipixuna, and Babaquara—that would have had twice the installed generating capacity as Belo Monte and together would have rivaled all existing dams in size. While the Xingu dams would have generated tremendous amounts of energy, they also would have forever changed the social and environmental landscape of the region. Kararão alone would have flooded an estimated 1,225 square kilometers, which is about three times that of Belo Monte, and the full complex of six dams would have inundated more than 18,000 square kilometers. Those dams would have displaced seven thousand Indigenous people and tens of thousands of others.[27]

The military regime was unable to begin the project before democratic governance returned to the country in the mid-1980s. In line with the country's downward trend of dam construction, subsequent administrations put the project aside because of other priorities, increasing national and international pressure to avoid large-scale dam construction, and, perhaps most importantly, the withdrawal of World Bank financing. The dams on the Xingu, however, were not completely abandoned. In fact, out of sight from the public's eye, technicians were redesigning the dam and studying its viability. In 1993, less than four years after the first proposal had been halted, officials from Eletrobrás and the Departamento Nacional de Águas e Energia Elétrica (DNAEE—National Department of Water and Electrical Energy) met to discuss how to proceed with the hydroelectric project on the Xingu. They decided that the project had to be redesigned with the sociopolitical viability of the project in mind. During the second half of the 1990s, a group made up of experts from Eletrobrás, Eletronorte, and DNAEE worked to update the plans in a way that would reduce the social and environmental impacts of the project.[28]

In 2002, Eletrobrás and Eletronorte released redesigned plans under the new name of Belo Monte, a gesture toward cultural sensitivity, as Indigenous groups did not want a word from their language used for a dam they did not want. These plans were nearly the same as those that would eventually be used to build the dam nearly a decade later. This redesign had drastically reduced the scope of the project. Instead of the six dams that were initially planned for the river, the Belo Monte facility would include two dams on the Xingu River and a canal that directed water from a smaller dam to the larger energy-generating dam downstream. Belo Monte would also be a run-of-the-river hydroelectric facility, which is a type of dam designed to reduce negative impacts by keeping just a small reservoir. Whereas traditional dams require storing large amounts of water and flood a significant amount of land, run-of-the-river dams like Belo Monte flood far less land and therefore displace fewer people. In the end, 478 square kilometers of land were inundated, a two-thirds reduction from plans that had emerged in the 1980s, and no Indigenous lands were directly flooded, which was also a significant change from the earlier plans. While close to forty thousand people would be displaced, half of whom were from rural areas and the other half from low-lying areas in the city of Altamira, many more would have been forced to move had the original plans been carried out.[29]

On the downside, the amount of electricity produced by a run-of-the-river dam fluctuates according to the amount of water in the river at any given moment, a consequence of rainfall amounts. For example, the installed capacity of Belo Monte is 11,233.1 megawatts. The average amount of "firm energy," or the energy guaranteed to be available, is 4,571 megawatts. Some people estimate that, during the driest times of the year, the dam will produce an average of just over one thousand megawatts. These inconsistencies, particularly in light of the increasing unpredictability of weather patterns due to climate change, are leading some decision makers to contemplate whether run-of-the-river designs make sense in future constructions.[30]

Belo Monte's design was particularly exciting for engineers. The complex was to be built in the Volta Grande (Big Bend) of the Xingu. The Volta Grande is a hundred-kilometer stretch of the river in which it curves approximately ninety degrees, turning from a northeast flow to a southeast flow before turning back toward the north. In this stretch, the elevation of the river drops over eighty meters. By comparison, the elevation of the Amazon descends a hundred meters from Iquitos, Peru, to the Atlantic Ocean, a distance of 3,600 kilometers. The significant elevation difference in the Volta Grande and the sharp curves in the river allow for a unique dam that theoretically has fewer environmental impacts. One dam is built in the Volta Grande, which diverts water into the manmade canal that carries the water to the other dam, located on the other side of the Volta Grande, which generates most of the complex's energy. An engineer who helped plan Belo Monte summarized the sentiment that the unique natural phenomenon was an ideal place for the dam when he reportedly said, "God only makes a place like Belo Monte once in a while. This place was made for a dam."[31]

DEMOCRATIC DEVELOPMENTALISM: A NEW APPROACH

Less than a year after the new plans were released in 2002, the Workers' Party—a left-of-center party that grew out of social movements—won the presidency and gained more control of the legislative branch. Anti-dam activists in the

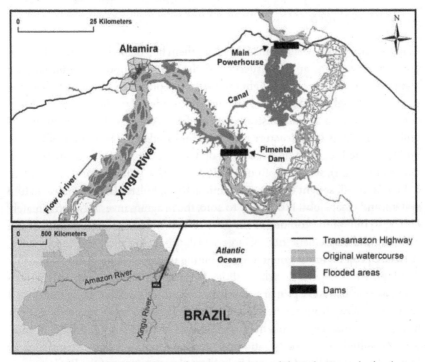

FIGURE 1.1. The Volta Grande region of the Xingu River and the Belo Monte hydroelectric facility. (Map created by Heather Randell.)

FIGURE 1.2. The main powerhouse dam of the Belo Monte facility, partially operational in 2017. (Photo by author.)

Transamazon region were thrilled. Many of them had helped strengthen the progressive movement, and the success of the party around the country, they assumed, should help in their fight for justice and put an end to plans to dam the river. Surprisingly, the Workers' Party, under President Luiz Inácio Lula da Silva, commonly known as Lula, instead instituted an aggressive development policy built around large infrastructure projects. Launched at the beginning of his second term, the Programa de Aceleração do Crescimento (PAC—Growth Acceleration Program), which continued as PAC-2 under Lula's successor, President Dilma Rousseff, sought to spur economic activity, bolster Brazil's infrastructure system, and create jobs. From 2007 to 2010, the program invested approximately USD$250 billion to promote, as the government stated, "the resumption of the planning and implementation of large social, urban, logistics, and energy infrastructure projects in the country, contributing to its accelerated and sustainable development."[32] Hydroelectric facilities became key components of the program, as one of PAC's primary objectives was to ensure that the country had sufficient energy from renewable and clean sources to meet its needs. Proponents argued that dams could help avoid crises like the energy shortages of 2001, which resulted in electricity rationing for nine months.[33] Large dams such as Jirau, in the state of Rondônia, and Belo Monte became icons of the PAC program and the overall approach to development that the federal government took under presidents Lula and Rousseff.

With the focus on large-scale projects and hydroelectric facilities, Lula's agenda was reminiscent of previous approaches to development, but as Congressman Geraldo referenced at the event commemorating the start of construction, the Workers' Party insisted their approach would be different from projects of the past. Instead of disregarding social and environmental concerns, proponents from the left, like Geraldo and Lula, contended that large-scale projects like Belo Monte could be enacted in better, more sustainable ways than in the past. A new era of dams, roads, and other large projects would not only contribute to economic growth and the needs of the country, they argued, but new participatory decision-making practices and extensive mitigation programs would reduce the negative impacts *and* bring necessary regional development and infrastructure to the places where these projects were built. Proponents of the approach suggested that these projects would spur economic growth, provide new opportunities, and offer better lives to people most directly impacted by construction.

In order to accomplish their goals, a great deal of financial resources would be directed toward managing the effects of large projects and strengthening the local infrastructure through the construction of new schools, medical facilities, and sewer systems, among other areas of need. The government would also institute forums for decision-making that would include everyday citizens, civil society representatives, and government officials. This openness to deliberation was a significant break from past developmental eras, which largely discounted local needs and visions for the future. By purportedly attempting to guide development through infrastructure growth, address inequality and poverty, *and* deepen democratic modes of governance, President Lula ushered in Brazil's version of democratic developmentalism.[34]

This approach to development aligned with the progressive government's ongoing attempts to demonstrate that they were an ally of the poor and disenfranchised. On a national scale, the Workers' Party was expanding its social programs and income redistribution policies, such as Bolsa Família, a cash-transfer program that started during President Cardoso's administration, and eventually Brasil sem Miséria, President Rouseff's attempt to rid the country of destitute poverty.[35] The party also expanded participatory opportunities throughout the country. Participatory budgeting, an initiative that allows the public to allocate portions of a municipal budget, spread from its first implementation in Porto Alegre in 1989 to thousands of cities around the county.[36] Management councils—which bring citizens together with government officials and service providers to develop public policies regarding health services, education, and the environment—also achieved widespread adoption. Sweeping reforms in the health and social assistance sectors upended bureaucratic and political networks, allowing for participatory institutions to take root nationwide. By the mid-2000s, over twenty-eight thousand councils had been established, and by a decade later, close to 100 percent of state and municipal governments in Brazil observed the mandate to implement health and social assistance participatory programs.[37] The federal government under Workers' Party administrations created new participatory initiatives, but these programs tended to be less meaningful than those started at the local and regional levels. Nevertheless, public participation expanded during the Lula and Rousseff presidencies.[38]

As participatory budgeting and management councils expanded throughout the country, the Workers' Party government used Amazonia as a laboratory for the incorporation of participatory opportunities in sustainable development initiatives. Within the first year of Lula's presidency, the administration launched the Plano Amazônia Sustentável (PAS—Sustainable Amazon Plan), which served as the guiding framework for regional sustainable development projects.[39] The first such project was to pave the Cuiabá-Santarém Highway, which opened during the military dictatorship, and create participatory development initiatives for communities who lived along the highway. Officials who worked on this project told me that there was a great deal of political will to carry out the plan in a participatory manner, but the scale of the project made it difficult to institute consistent and inclusive opportunities. There was also little follow-through, leaving the project largely unfinished and participatory promises unfulfilled.[40] Despite the disappointment, some of the officials involved in the project wanted to build on these ideas, so they utilized a more participatory process in creating a development plan for the Marajó island in the state of Pará in 2006–2007.

Given the attention to social programs and participatory governance, it is no surprise that the government invested a great deal in these efforts alongside Belo Monte, the largest construction project in the country. As a result, Belo Monte soon became the government's signature project, representing the purportedly more sustainable, inclusive, and just approach to development. Proponents celebrated the dam as a way to generate substantial clean energy while also bringing growth and opportunities to the people of the Transamazon, a region that had been historically neglected. There were tangible resources behind the rhetoric and provisions designed to protect the local people and natural environment. The licenses to build and operate the dam were conditional on Norte Energia meeting dozens of guidelines meant to mitigate negative social and environmental impacts from the dam while also improving infrastructure and bringing economic growth to the region. In order to meet these requirements, nearly four billion Brazilian reals (approximately two billion U.S. dollars at the time of construction), of the thirty billion total cost of the project, were dedicated to mitigation efforts, sustainability initiatives, and local infrastructure projects. Of these four billion reals, over three billion were invested in predesigned mitigation programs laid out in the *Projeto Básico Ambiental* (*PBA—Basic Environmental Project*), which was drafted by consultancy companies. In its seven volumes and more than 2,500 pages, the *PBA* describes, in great detail, the plans for addressing environmental and social impacts of Belo Monte.

The government also created a number of participatory forums that gathered government officials, nongovernmental organizations, community members from dam-affected areas, and Norte Energia representatives. Some of these spaces were for information sharing while others had real decision-making power. Arguably the most significant of these participatory opportunities was the Comitê Gestor do Plano de Desenvolvimento Regional Sustentável do Xingu (CGDEX—Steering Committee of the Regional Sustainable Development Plan of the Xingu), a participatory council focused on bringing "sustainable development," as defined by the participants, to the region. President Lula initiated CGDEX when he signed a decree in 2010 that laid out the basic organization and purpose of the committee.[41] The members of the committee—which were federal, state, and municipal government officials and regional civil society representatives—met monthly and welcomed anybody to participate. They were tasked with allocating 500 million Brazilian reals (approximately 250 million U.S. dollars at the time)—an amount the government mandated Norte Energia dedicate to the committee—toward regional development programs. Participants debated what they should include as "sustainable development," ultimately using broad understanding of the term and funding a range of projects, from limited arts-based cultural initiatives to supplementing necessary health care needs in the region and much more. The committee had allocated approximately half of the total funds when right-wing president Jair Bolsonaro terminated the project in 2019, less than six months after he took office.

While CGDEX had real decision-making power and was at the center of a great deal of conflict, it was not the only forum for dialogue and information sharing. For example, licensing processes required that Norte Energia hold *audiências públicas*, or public hearings, prior to constructions. At these hearings, Norte Energia representatives explained the upcoming projects and citizens could voice concerns. During construction, Norte Energia also had to institute the Fórum de Acompanhamento Social, or Social Monitoring Forum—a series of meetings in which the construction consortium and its subcontractors shared information and updates with community representatives about the social programs that accompanied the dam. While only specific members were invited to participate, others received invitations to certain meetings or gained access to the forum after raising awareness of their issues. For example, the fishers gained access to the forum as part of the agreement to end the occupation after their monthlong protest.

Norte Energia also held dozens of community meetings specifically related to the resettlement of residents who would be displaced by rising waters from the dam's reservoir. These meetings took place in the neighborhoods people would be forced to leave, in order to share the compensation options and the plans for new neighborhoods and homes that Norte Energia would build for them. Norte Energia used the opportunity to recruit neighborhood representatives to serve on a housing council that would accompany the resettlement process. These representatives were responsible for relaying information about the process to their neighborhoods.

Proponents of the dam argued that these initiatives would curb any negative impacts and bring regional development—including economic opportunities, better education and health services, and other livelihood improvements—to an area that the government had largely ignored since the 1970s. Many advocates, in fact, used the mitigation plans and participatory opportunities to turn opposition arguments on their head, insisting that what appeared to be drawbacks could spur growth and improvements in livelihoods. For example, the government argued that the well-being of most displaced people would improve. Many urban residents lived in *palafitas*, hand-built "stilt" houses elevated to accommodate the rise and fall of river levels. During my time in Altamira, I repeatedly heard government officials and Norte Energia representatives argue that people in these communities were living in subhuman conditions and, as a result, would benefit greatly from dam's construction. I did find a few people in those neighborhoods who were excited about moving to a new home. In rural areas, where residents were displaced well before urban areas, research conducted shortly after people were displaced suggested that the well-being of those families did, on average, improve.[42]

FROM RHETORIC TO IMPLEMENTATION

The actual implementation of the democratic developmental approach was predictably complicated. Belo Monte was a massive project that, like previous dams, caused both anticipated and unforeseen social and environmental side effects. These impacts deepened the conflict and tension that surrounded its construction, despite—and sometimes because of—the attempts to manage those negative effects and to bring more people into decision-making processes. Furthermore, while the massive investment in mitigation and participatory programs allowed the state to argue that it was bettering life for everybody in the region, the programs that the government mandated Norte Energia implement were often delayed, not useful, or led to unexpected problems.

When it came to resettlement, most residents lacked clear information about the process. Prior to the displacement, and for much of my time in Altamira, people were uncertain about the impacts of the rising waters and the fate of those who would be forced to move. The compensation program for displaced residents and business owners was detailed yet confusing. Many people did not know what option best served their interests, nor did they understand what roles they could play in the negotiation process. While some families were relieved to have new houses, others were unhappy they would have to leave their homes. After people moved to the newly built Reassentamentos Urbanos Coletivos (RUCs—Urban Resettlement Collectives), a host of problems ensued. Most of the new neighborhoods were located far from the city center. Many people did not own a car or motorcycle, and the city did not have a public transportation system for over two years after the major period of resettlement. By the end of 2017, some of the communities were experiencing severe water shortages. Furthermore, the influx of people and money to the region had also brought skyrocketing drug trafficking and crime-related violence, much of it in the resettlement communities. By 2015, in fact, Altamira had become the most violent city in Brazil, as measured by the homicide rate, a frightening fact given that some of the country's large cities are among the most dangerous in the world. People living in the resettlement communities bore the brunt of this violence. These dynamics made housing and displacement a major point of conflict and reason for dissent, deliberation, and claims making.[43]

The scale of Belo Monte also meant that there was rapid and uncontrollable population growth, which led to a host of other problems. The population, by some accounts, had increased from less than 100,000 in 2010 to 150,000 just two years later and had nearly doubled by 2014.[44] Thousands of people migrated to Altamira looking for work—as nearly thirty-five thousand new jobs would be available by the peak of construction—and many brought their families. Others knew that this population increase meant there would be needs for other businesses, particularly in the food and hospitality industry. Existing owners of hotels and restaurants quickly remodeled and upgraded their facilities, while others built anew, leading to a sudden proliferation of available construction jobs and an increase in service work.

While these dynamics led to rising employment, the resulting population explosion caused a skyrocketing cost of living, led to the aforementioned violence, increased pollution, and strained already overextended infrastructure, such as the education and healthcare systems. Rent prices and hotel costs more than tripled in the center of the city, even as new lodging options opened on the outskirts of town. Additionally, some of the new arrivals to the region were what are sometimes called *barrageiros* in Brazil, or those who travel from one dam site to another for work. Some specialize in certain aspects of dam construction or management, while others set up temporary small eateries or lodging options near construction sites. Many *barrageiros*, who are predominantly men, often live in temporary communities. These communities increase both voluntary and involuntary prostitution, as well as drug-trafficking operations. Police uncovered one prostitution and drug-trafficking ring in 2013 near the Belo Monte dam.[45]

The population boom also led to increased pollution in various forms. Air pollution worsened as a result of a sudden rise in the number of vehicles on the road. Most significantly, hundreds of buses transported thousands of workers from Altamira to the construction sites at all hours of the day. Not only did the exhaust deteriorate air quality, but the buses also kicked up a great deal of dust. Water pollution also worsened. The lack of a city sewer system meant that more and more waste ended up on the streets and in the river. The already overcrowded clinics and hospitals were further strained, leaving people with long wait times, and, according to anecdotal reports, some people failed to receive necessary care.[46]

The fact that the anticipated and familiar negative impacts of dam construction came alongside a glut of resources and promises for regional development that accompanied Belo Monte presented particular challenges and opportunities for the people affected by the dam and their advocates. As in other places where dams had damaging consequences, many of the people adversely impacted did not simply accept the harmful effects. Rather, they used a host of strategies and structures to demand a different outcome. How they did so, and the difficulties they faced, had much to do with previous centuries of exploitation in the region, as well as the storied local history of collective organizing since the 1970s.

2 · BOOMS, BUSTS, AND COLLECTIVE MOBILIZATION ALONG THE TRANSAMAZON

During the construction of Belo Monte, Antonia Melo da Silva could be found at nearly every anti-dam rally and protest for rights and recognition in Altamira. She was often speaking forcefully to a crowd or a group of journalists about the injustices associated with the dam and large-scale development. Her cutting speeches, critiques of the government, and seemingly endless energy to continue fighting belied the fact that when construction on the dam began, she was in her mid-sixties and had been an activist along the Transamazon for over four decades. In that time, Antonia had become the face of dam opposition, the women's movement, and the broad struggles for social justice in the region. In 2008, she helped found the Movimento Xingu Vivo Para Sempre (Xingu Vivo—Xingu Forever Alive Movement), which remained the most outspoken and staunch anti-dam organization throughout construction as they supported fishers, riverine people, farmers, Indigenous people, and other dam-affected communities.

Antonia was born in the northeast state of Piauí in the late 1940s and moved to Altamira with her family when she was four years old. They settled on a rural property near the city, where they raised small numbers of chickens, ducks, and cows and farmed coffee, cacao, rice, and beans for both consumption and sale. There were few schools, health care facilities, or other basic services. Antonia married in 1970, built a new house near the center of the city, had four children, and became a teacher. Her work as an activist began on her street, where she helped organize residents to demand that authorities provide water and energy. In the 1980s, Antonia became involved in the Catholic Church's efforts to mobilize residents, and as a self-defined militant, she helped the Workers' Party grow in the Xingu region. Through the organizational power of the church and the political vehicle of the Workers' Party, she marshaled residents to protest the first plans to dam the Xingu and make demands on the government for better schools, health care, and roads. She was integral in the creation and work of regional social movements and nonprofit groups that residents credit for the construction of the regional hospital, the birth of the university in Altamira, road improvements, and access to credit for farmers. The mechanisms through which Antonia became an engaged participant in social, civic, and political life in and around Altamira are representative of many of the people who would eventually be active in the opposition to the construction of Belo Monte.

The lack of infrastructure, government support, and access to economic opportunities made life challenging for Antonia and most people in the region. A long history of outsiders exploiting the rich natural resources of the Xingu basin and much of Amazonia—yet largely disregarding the people living there—resulted in these difficult conditions. Since Portuguese explorers first arrived in South America in the 1500s, politicians and businesspeople have repeatedly sought to profit from the region's resources, yet have provided scant benefits for local populations, offered few environmental protections, and often deserted projects shortly after reaping as many gains as possible. The intermittent extraction of resources created a series of boom-and-bust cycles, in which a great deal of opportunities and resources flowed to the region, only to come to an end a short time later. These cycles intensified in the 1970s with the military dictatorship in Brazil, which launched but failed to fully implement a process of frontier expansion into Amazonia, including dam construction, policies to encourage economic investment, road construction projects, and a settler colonization program.[1]

The repeated exploitation and abandonment led to instability for the people in the Xingu region, but also created the conditions through which an active and unified civil society emerged at the end of the twentieth century. Each time the state and businesspeople abandoned the region, people like Antonia worked with and strengthened various institutions and organizations to fill the gaps. The Catholic Church had a long presence in and around Altamira, often working to provide infrastructure and organize local residents. By the 1970s and 1980s, labor unions, social movements, women's groups, and the Partido dos Trabalhadores (Workers' Party) joined the church in making demands on the state to improve the infrastructure in the region and resist the construction of dams on the Xingu River. The colonial history of the region, liberation theology, the negative impacts of previous dams, the government's failed developmental interventions, the struggle for democracy throughout the country, and the growing strength of the Workers' Party and labor unions unified civil society and made protest a powerful tool in the region from the 1970s to the first decade of the twenty-first century.

FROM EXPLORATION TO EXTRACTIVISM

From the early seventeenth century to the late nineteenth century, slavery, missionary, and colonial efforts brought Europeans inland to create settlements in what is now known as Amazonia. As the early explorers needed assistance navigating the waterways and dense jungle, they forced Indigenous populations into slavery. Jesuit missionaries followed on the heels of explorers and colonists, serving both to protect Indigenous populations from slavers and to transform the material and cultural life of these groups. This early migration of colonists, slavers, and missionaries fundamentally altered many Indigenous groups' ways of life throughout the Amazon and decimated native populations when the outsiders introduced European diseases against which Indigenous people lacked immunity.[2]

Other than small settlements, migration to the Xingu region was minimal until the 1880s, when the region became central to the Amazon's rubber boom. The world demand for rubber skyrocketed, first due to the bicycle craze in Europe and the United States and then to the automobile industry. For many years, the Amazon was the sole source of the raw material for tires, and the invention of the steamship eased the costs of exporting the product from the forest to foreign markets. These global dynamics, coupled with a drought and the end of the cotton boom in the northeast of Brazil, enticed intrepid Brazilians to venture westward into Amazonia, in hopes of striking it rich through rubber tapping. As the market expanded, rubber barons and slaveowners from the northeast built towns throughout the region in efforts to establish and maintain power. One such landowner was Raymundo José de Souza Gayoso. Like others, Gayoso's large coffee plantations in the northeast were failing, so he brought his slaves and resources to the Xingu region in 1883, in hopes of producing manioc flour and rubber. His slaves opened up an important road from downstream in the river to what is now Altamira. The road made it possible to move through the region without boating the Volta Grande of the river, which was complicated and sometimes impossible to navigate, given the number of waterfalls and rock formations. After Brazil abolished slavery in 1888, Gayoso returned to the northeast, but other explorers, landowners from the northeast, and rubber tappers followed the path opened by Gayoso's slaves, settling in and around Altamira. The population and economic growth led to the formal incorporation of Altamira as a municipality in 1911, around the time the rubber boom peaked.[3]

Shortly thereafter, the African and Asian production of rubber became more efficient, leading to a crash in the Brazilian rubber industry. By 1920, revenues from rubber all but disappeared, national and international interests began turning elsewhere, and the Xingu region was left without a local economic industry to support the population.[4] With little infrastructure, lack of investment in anything other than rubber, and physical isolation from the economic and political centers of Brazil, Altamira went into a period of economic stagnation. Given the limited opportunities, some residents moved away, and few people moved into the region, leaving it sparsely populated for much of the twentieth century. According to the 1960 census, fewer than twelve thousand people inhabited Altamira, which by land area was the largest municipality in Brazil, covering an expansive sixty-two thousand square miles.[5] Those who stayed took up other extractive trades such as hunting, fishing, and gathering. As the economy shifted, Brazil nuts, animal pelts, natural oils, and other commodities from the Amazon expanded the export market, and the region became more monetized, even as the federal government invested very little in the area.[6]

In the absence of the Brazilian government in the region, the Catholic Church became a principal driver of local development from the 1930s to the 1960s. In the Xingu region, church authorities worried that, in the wake of the rubber boom, the lack of educational opportunities, health services, and social assistance would lead to the exodus of both the rural and urban populations, so they began investing in local infrastructure. In addition to constructing the first cathedral in Altamira in 1939, they built the first primary schools shortly thereafter. The church then created a center of support for rural families, which offered a small carpentry shop as well as a few small machines to convert raw materials into usable manioc flour, sugar, honey, and rice. In 1953, they opened a secondary school and subsequently raised money to build the first regional hospital, which opened in 1967. The church's work during this period provided the basis of the organized and strong civil society—led by people like Antonia Melo—that would come to define the area during and after the military dictatorship.[7]

Meanwhile, other parts of Brazil changed between the 1920s and 1960s in ways that would have significant impacts on the Amazon, the Xingu basin, and Altamira. It was during this time that President Getúlio Vargas was supporting industrialization in the south of the country and beginning to build Brazil's developmental state. This industrialization would drive up the demand for raw materials, many of which would be found in the Amazon. During the same period, the government also actively expanded its presence beyond the larger metropolitan areas of Rio de Janeiro and São Paulo. Most significantly, in 1960, the capital moved from Rio de Janeiro to Brasília, a newly constructed city in the center of the country. These dynamics set the stage for military expansion into the Amazon in the 1960s and 1970s.[8]

COLONIZING THE AMAZON

Soon after the Brazilian military seized control of the government in 1964, leaders looked to the Amazon for economic and geopolitical security. The military dictatorship first favored foreign capital and economic restructuring, but their political base of support, the national bourgeoisie, realized they would not benefit from these initial development plans. In part to satisfy elite interests, the Amazon soon became central to the new government's political and economic agenda. Beginning in 1966, officials instituted policies that would spur private investment in the Amazon and created an administrative agency to manage programs in the region. The military leaders saw the region as an ideal place to expand Brazilian-based investment and industrialization policies, given the widely held view that Amazonia had extensive amounts of untapped resources and sparsely populated land. Geopolitically, the government viewed the north as a vulnerable border region, so bringing people, industry, and infrastructure projects could help secure the country from invasion. In addition, frontier expansion aligned with nationalist rhetoric of the military and the country's elites. As Charles Wood and Marianne Schmink summarize, "The 'conquest of Amazonia' became an exalted mission linked to the grandeur of the nation itself."[9]

One of the most consequential initiatives in this goal of conquering Amazonia was the encouragement of new settlement in the region. The state's professed purpose of such settlement would be to relieve population pressures in the northeast and provide land—which the government claimed was uninhabited—to poor citizens. In 1970, the government launched the Programa de Integração Nacional (PIN—National Integration Plan), an ambitious infrastructure and colonization program that would fundamentally alter the social, economic, and environmental landscape. PIN provided funding for the construction of the Transamazon highway, which was designed to cut through the dense forest and stretch nearly three thousand miles from the Atlantic Ocean to the western edge of the country. Settlers were to be given hundred-hectare plots of land located along ten-kilometer-long roads on either side of the highway. In a departure from previous policies, which supported private capital investment and large business, PIN intensified federal activity in the region and focused on small-farmer colonization. The government argued that the focus on infrastructure development through road construction and colonization in the Amazon was an attempt to relieve pressures facing the overpopulated and poverty-stricken northeast that was experiencing severe droughts. They promoted the area for potential settlers as a "land without people for a people without land"—a clear example of the disregard for the many different groups already living throughout the Amazon. Most scholars argue, however, that the extensive mineral deposits, land for agricultural growth, lumber, and the ongoing geopolitical concerns were significant factors driving the highway project.[10]

Until the construction of Belo Monte, the Transamazon and colonization project had the most significant impacts on Altamira than did any other policy or program. Centrally located along the new highway, the city served as the nucleus of colonization activity and was intended, over the long term, to be used as a supply center for the smaller planned communities to the east and west. Altamira was also the symbolic center of the government's integration and expansion efforts, evidenced by President Médici's visit to mark the beginning of construction in

1970. The formal colonization process, the government's incentives for private investment, and the new business and industry opportunities that came with the construction of the Transamazon led to a rapid population increase in the region, particularly in Altamira. Thousands of people from the northeast, most of whom had few economic resources, took advantage of the free land the government offered. The economic opportunities during this time also contributed to the in-migration of wealthier people from the south of Brazil. In approximately five years, the population doubled in the urban area of Altamira, from about six thousand at the end of the 1960s to about twelve thousand in 1975. As a result, the basic infrastructure of the city improved considerably. Hundreds of new commercial businesses opened, phone service was installed, electricity arrived, municipal markets opened, some streets were paved, and the airport expanded to accommodate jets.[11]

While some services improved, the centralized nature of government decision-making and the military's focus on national economic growth and its own political needs meant that the dictatorship's policies and programs fundamentally supported large-scale agriculture, cattle ranching, and industry. The government essentially saw poor and small-scale farmers, miners, native people, and the natural environment as obstacles to, rather than the focus of, development. The roads may have opened up the region and connected it to other parts of Brazil, shifting the transportation and communication from the waterways to the roads, but the government never developed a plan for how to support the local population. The soils and climate were unsuitable for annual crops, the high costs of road construction and maintenance meant little buy-in from politically powerful groups in the region, and Brazil's ongoing dependence on foreign capital mixed with the rise in oil prices in 1973 to create economic challenges for the country. These factors led to the government's focus on rapid economic gains instead of addressing social welfare goals. Furthermore, the government's attention to the colonization program was short-lived. Soon after the majority of highway construction was complete, the state returned to policies that encouraged development through large-scale enterprises, and the government abandoned the settlement projects it had just begun. The military dictatorship's nationalist goals of expansion, its inability to control the frontier, the lack of government foresight and follow-through, and its politically driven policies left the people of the region to fend for themselves and the poorest struggling to survive.[12]

SURVIVAL AND THE CATHOLIC CHURCH

Simone came to the Transamazon Highway with her family in 1978, toward the end of the colonization period. She was a young child at the time, but decades later could vividly recount the experience of first arriving on a rural lot in the middle of the forest: "I remember the dense immense forest, the bugs and animals, the big rivers and such." Simone's family was moving to the abandoned lot of a family that had moved to the area earlier that decade but had already left because at least two of their children had died from malaria. In order to reach the lot, Simone, along with her parents and siblings, had to walk approximately five kilometers from the Transamazon highway. Simone recalled her mother's reaction: "My mother was horrified when my father stopped at a straw tent and said, 'We've arrived!' She asked, 'Arrived where?' 'This is our house.' 'Here? We are going to live here?' she exclaimed. My mother spent a week crying. The only reason she did not leave immediately was because she did not know which way to go."

As settlers like Simone's family arrived, they faced immense difficulties, despite government promises of new opportunities. Small farmers were economically, technically, socially, and politically ill-equipped to adapt to the region. The climate and soils were quite different from where they originated, which made farming challenging, and they encountered a host of wildlife, from intense mosquitos to snakes to large predators. Settlers struggled against disease and hunger. The government failed to maintain the roads, making them impassable during the rainy season, which could last for six months at a time. This made reaching the few health posts and schools difficult, and often impossible, for many families. Given these extraordinary challenges, the state's absence, the lack of infrastructure, and the difficult conditions as people rushed to the region in the 1970s, new migrants like Simone and longtime residents like Antonia Melo organized in efforts to improve their conditions. This collective organizing would eventually lead to a unified network of individuals, organizations, and social movements to make demands on the state, fight against dam construction, and work to support the people who lived in the region.

Early struggles to improve conditions in the region were rooted in, and empowered by, ideas of liberation theology. Liberation theology draws on Marxist understandings of class conflict and suggests that the Christian faith calls believers to work for and alongside marginalized communities. As Leonardo Boff and Clodovis Boff, two of the earliest and best-known Brazilian supporters of this approach to faith and social justice, wrote, "Liberation theology was born when faith confronted the injustice done to the poor."[13] Some of the prominent liberation theologians used Marxism—which was fast becoming a dominant worldview and ideological paradigm for activists throughout Latin America—to identify specific causes of poverty and radical ways to eliminate it. Progressive sectors of the church became instrumental in mobilizing and raising the political awareness of marginalized communities, which helped spread the ideas of liberation theology throughout the continent in the 1960s and 1970s. The priests and church members who were politically oriented toward liberation theology often used what are referred to in Brazil as Comunidades Eclesiais de Base (CEBs—Christian Base Communities) to organize residents. CEBs often created a religious experience that focused on building community solidarity and discussing residents' lived experiences, rather than on accepting church authority or doctrine outright.[14]

Liberation theology also influenced the Conferência Nacional dos Bispos do Brasil (CNBB—National Conference of Brazilian Bishops) to create the Conselho Indigenista Missionário (CIMI—Indigenous Missionary Council) in 1972. CIMI has since advocated for the rights of Indigenous people in Brazil, helped create and support demarcated Indigenous territories, and defended the rights of Indigenous communities facing violence, injustice, invasion, and other forms of domination. The group has helped organize Indigenous communities and their advocates at regional, national, and international scales, and they have used their linkages with the Catholic Church to influence decision makers.[15]

In the Xingu region, liberation theology, the use of CEBs, and CIMI were fundamental to claims making during the dictatorship. The bishop Dom Erwin Kräutler and the *prelazia* (territorial prelature) that he chaired from 1971 to 2015 were important focal points for the region's activism.[16] Kräutler arrived in Altamira in 1965 as a young priest from Austria. He taught primary school for fifteen years, until he became the bishop representing the region. He also served as president of CIMI from 1983 to 1991 and again from 2006 to 2015. Kräutler used his position in the church to fight for the rights and livelihoods of the poor and Indigenous and against the type of development pushed by the military dictatorship. By the time dam construction had begun, he had become such a recognized and outspoken activist that he received death threats from opponents to his activism. He traveled with personal bodyguards everywhere he went, and reports suggest that he often wore a bulletproof vest.

In addition to the increasing attention to Indigenous needs, the prelazia focused much of their work in the 1970s and 1980s on supporting the struggling settlers. A small number of priests would venture out along the Transamazon to create CEBs and talk with people about the difficult living conditions and what the church could do to support them. As a result, the churches along the highway became the reference point for meetings and mobilization efforts for those who stayed on their land. Due to the harsh living conditions, other settlers abandoned their land and moved to the urban center of Altamira. Many of those settlers had little to no resources, so the prelazia allowed them to build houses on the urban land that the church owned. The church was thus an active participant in the growth of both rural and urban areas, as well as a key avenue through which people sought better living and working conditions.

From the 1970s to the 2010s, nearly every civic leader in Altamira, including Antonia and Simone, and many elected officials in the region started their political engagement, social organizing, and protesting through the Catholic Church. For at least fifteen years in the 1970s and 1980s, the church taught people why and how to organize members of their communities. Antonia's work as a coordinator for the CEBs in the late 1970s showed her that people have to fight for rights when those rights are not respected. She did not believe that God worked miracles to solve social problems; rather, the church taught her that scripture called on people to come together to bring about social change. The notion of working together to demand rights and create change has remained a fundamental belief that guides her activism and the civic engagement of many others.

Similarly, the Catholic Church became a central part of Simone's life as she and her family faced the challenges of new migrants. Along with their neighbors, her family helped build a small church, which served as a place of worship, a school, and a meeting spot. Her mother became a teacher, and even at a young age, Simone was inspired to help improve living conditions for her family and neighbors. A priest from the region soon invited her to participate in a group of engaged young people outside of her community. "From there, it never stopped," she summarized. She went on to talk about the decades of work she had done to fight for rights and justice in the region, organizing marches, meeting with community members regularly, and demanding that city governments in the region build schools, hospitals, and other basic infrastructure. Her work through the church led her to be a self-defined nonviolent militant. The church's efforts along the Transamazon not only built the leadership capacity of individuals like Antonia and Simone; it also created a unified group of engaged citizens who together fought for a better way of life in the region, which led to social movements and other nongovernmental organizations.

The organizational work of the church, along with the closely affiliated workers' unions, led to the first region-wide protests aimed to raise awareness about and improve poor conditions. In 1978, farmers blocked the Transamazon highway to show their frustration with the prices of agricultural products.[17] Given that the Transamazon was, and still is, the only land-based transportation route in the region, closing access to the highway effectively drew the attention of government officials and other residents. Protestors have repeated this type of action many times since. In May 1983, for example, a group of sugarcane workers blocked the highway for more than ten days to protest not receiving wages. The protest eventually ended when the military police arrested some of the protesters after using tear gas and stun grenades, but the workers' employers eventually paid the overdue salaries. Among those arrested was Bishop Kräutler, who later considered the protest to be "the start of everything," because it was based on "action, autonomy, and nonviolence."[18]

Collective organizing and social movement activity increased over the subsequent decades, with rallies and protests around a wide assortment of issues. Many of the people who have since become outspoken activists on a number of issues began their engagement by raising awareness about and fighting for three basic needs: education, health, and transportation. Mika, a migrant to the region during the colonization project, started a movement for improving education and better conditions for rural farming families in the late 1970s. As a teacher, she saw the state's lack of investment in education, particularly in the rural areas. Few schools existed in rural areas, and those in urban centers like Altamira lacked resources, including sufficient teachers. "Sure, many adolescents traveled to Altamira for school," Mika remarked, "but we needed schools in rural areas to improve the viability of agriculture in the region." She gained experience through her work with social movements, the Catholic Church, and the rural workers' unions. These experiences gave Mika the knowledge to form and become coordinator of a union dedicated to improving conditions for rural teachers, students, and education in general. With the help of other engaged citizens, she held meetings, protests, and rallies to raise awareness of the dearth of resources and called on government to construct schools in rural areas. Participants credit their direct-action techniques with bringing new schools to the region in the 1980s–2000s, and the union remained active through the construction of Belo Monte.

The lack of quality roads was also a significant challenge as settlers arrived. Most lived on farms away from town centers like Altamira and basic services such as schools and health clinics. Families found it difficult to navigate the dirt roads in the best conditions, and the seasonal rains turned those roads into mud, isolating families from basic services. During the rainy season, even the Transamazon could be impassable for months at a time. These poor transportation conditions made rural organizing difficult, but they also provided an issue around which many people would rally because they affected everyone. The inadequate roads made increasing the number of schools and health posts essential for residents and activists.

Many of the activists who worked to improve conditions in rural areas were also able to draw attention to similar issues in the urban center of Altamira. Mika and other members credited their union, as well as broader organizing and protesting, with bringing a new school and large sporting space to the city, which have been used by an assortment of residents for sporting, other large-scale events, meetings, and gatherings for decades. In addition, residents and university professors in Altamira often attribute the establishment of higher education in Altamira to the social movements. In the broader fight for increased education, the regional social movements demanded that the state offer college courses to train teachers. They argued that this would supply the region with competent teachers *from* the region who could fill the increasing number of schools. In 1987, at least in part due to their demands, a campus of the federal university, Universidade Federal do Pará (UFPA), opened in Altamira. At the time, it offered courses in areas that could train teachers: pedagogy, history, mathematics, language, and geography.[19]

The experiences of Antonia Melo, Simone, and Mika are common among the local civil society leaders. They were grounded in the church's social mobilization of the 1970s and 1980s, which provided unity among those fighting for rights. The church's activities highlighted the effectiveness of community organizing and protest and gave residents the tools to be active participants in the region's social and political landscape. This foundation would prove important during Brazil's transition to democracy and as the initial plans to dam the Xingu emerged.

HALTING DAMS ON THE XINGU RIVER

In the late 1980s, the collective mobilization that the Catholic Church helped catalyze along the Transamazon provided a strong network that activists could use to mount opposition to proposals for dam construction on the Xingu. The national and transnational anti-dam movement had also grown throughout the decade, providing broad support for local resistance efforts. The military regime officially ended in 1985, and as the country transitioned to democracy, the anti-dam movement found allies in the broader struggle for social and environmental justice. Additionally, Indigenous communities around Brazil—with the support of anthropologists, the church-linked CIMI, and activists—had been organizing politically for more than a decade. Their efforts to control their land and culture while fighting against illegal mining, foresting, and other extractive practices made them natural allies with others who opposed hydroelectric facilities.[20] As a result, civil society resistance was strong in 1987, when Eletrobrás, the state-run electricity company, released its *Plano 2010* (2010 Plan), which detailed plans to construct 297 dams in subsequent decades. The plans to dam the Xingu, a key part of the *2010 Plan*, had been leaked earlier that same year, giving time for organized resistance to grow.[21] Given the scale of impacts and the number of people the dams would affect, local and international social movement alliances, environmental advocacy organizations, and Indigenous groups viewed these dams as far worse than those already under construction and planned. After the struggles for health care, education, and other basic needs along the Transamazon highway, the mechanisms for collective mobilization in the region were already in place. The stage was set for the possibility that activists could halt dam construction.

Shortly after Eletrobrás released the plans for dam construction, the Catholic Church, led by Bishop Kräutler, began holding community meetings to discuss the "true history of the hydroelectric project."[22] Church leaders were able to use their networks and trust to gather information from local workers who had accompanied the viability studies for the dam. Much of this information contradicted data in the official report. Church leaders then shared their findings with other residents, and these meetings laid the groundwork for later protests and activism against the dam. Other activist groups joined the church to rally against its construction.

As regional organizing against the dams on the Xingu grew, the resistance garnered a great deal of attention around the country and the world. Local Indigenous leaders, according to Bishop Kräulter, approached the church in 1988 to ask for support in organizing a large protest against the dam. In the February 1989

FIGURE 2.1. Tuíra Kayapó confronting Eletrobrás engineer José Antônio Muniz Lopes at the First Gathering of the Indigenous Nations of the Xingu, 1989. (Photo by Paulo Jares / Abril Comunicações S.A.)

event, the First Gathering of the Indigenous Nations of the Xingu, over six hundred Indigenous people from about forty regional tribes gathered to share their disdain for the proposed dams. The gathering received international attention, in part due to celebrity support and attendance by musician Sting and The Body Shop founder Anita Roddick, and in part due to a dramatic moment caught on camera by international media. Tuíra, an Indigenous Kayapó woman, confronted José Antônio Muniz Lopes, an engineer for (and eventual CEO of) Eletrobrás, as he was addressing a crowd of people. According to reports, she dramatically lowered a machete to within millimeters of Lopes's shoulder, pressed the flat part of the blade against his face, spat at him, and told him, "You are a liar. We don't need electricity. Electricity won't give us food. We need the rivers to flow freely: our future depends on it. We need our forests to hunt and gather in. We don't want your dam."[23] As the Indigenous community and international stars gathered to confront company and government officials over the negative effects the dams would have on Indigenous populations, other groups organized residents of Altamira to hold a concurrent public protest in which they argued that the government had deliberately excluded civil society from the decision-making processes over the dam. They publicly criticized the extensive effects the dam would have on the local population and the environment.

As these events were taking place in Altamira, the World Bank decided to withdraw the loan it had promised to Brazil for the project. Due to the lack of funding, the federal government halted the Kararão dams before construction began. Some scholars and activists attribute the World Bank's decision to the protests, while others suggest that the transnational movement was only part of the bank's motivation. At the time, the developmental state in Brazil, which promoted large government-led infrastructure projects as the norm, was shrinking, in large part due to structural conditions.[24] The Latin American debt crisis had struck in 1982. Countries could not repay loans to foreign investors, and hyperinflation remained a serious problem in Brazil until 1994. One consequence was that state-led firms like Eletrobrás could no longer receive loans for large infrastructure projects.[25] Meanwhile, Brazil's Banco Nacional de Desenvolvimento Econômico e Social (BNDES—National Bank for Economic and Social Development) was beginning to focus on export promotion rather than infrastructure development and was not in a position to fund the project.[26]

Additionally, the late 1980s marked the first years of a return to democratic governance in Brazil. New federal environmental licensing processes that took social and environmental impacts into consideration were enacted in 1986, and the new constitution took effect in 1988, all of which brought about new protections for traditional communities and the environment, along with increased public scrutiny of such projects.[27] The growing domestic attention to social and environmental issues in Brazil was part of an emerging global debate over how to balance economic development with environmental and social needs, or, in other words, how to achieve "sustainable development."[28] In this context, the World Bank faced increasing pressure from a vocal international civil society over the impacts of its projects, particularly large dams. It was not an ideal time for the bank to fund one of the largest infrastructure projects in the world, and dam construction on the Xingu stopped, at least temporarily.[29]

CIVIL SOCIETY CONSOLIDATION ALONG THE TRANSAMAZON

The fight against the dam further unified activists who were demanding basic rights and infrastructure in the region. The dam resistance efforts also helped locally engaged residents connect with national and transnational movements. On the heels of the anti-dam protests and with a growing network of support, collective mobilization continued to expand through the 1990s along the Transamazon. Both longtime residents and newcomers wanted to influence how development would be carried out and address pressing local social and environmental issues.[30] In order to do so, engaged residents, particularly women, in the Altamira region built upon the experiences of collective organizing through the Catholic Church and consolidated their struggles for social justice and human rights through new social movements and the strengthening Workers' Party.

Women began coming together to address gender-specific issues. One of the most pressing problems was domestic violence, which had become common since construction of the Transamazon began in the early 1970s. The rates at which women were suffering and dying from violence carried out by partners were exceptionally high, but there was no place for women to turn to when experiencing abuse. In order to confront this and other gender-specific challenges, a group of women who had been active in the Catholic Church's CEBs formed the Movimento de Mulheres Trabalhadoras de Altamira—Campo e Cidade (MMTA-CC—Movement of Women Workers of Altamira—Country and City) in 1991.[31]

Through MMTA-CC, women utilized a variety of approaches to support other women and children. Maria Silva, a longtime engaged resident, professor, and coordinator of the Altamira campus of the federal university, describes the group's activities in her doctoral thesis about migrant women along the Transamazon. She explains that the MMTA-CC held marches, officially denounced acts of violence and abuse, accompanied legal processes, publicly demanded that criminals be punished, and held assemblies and presentations to discuss public policy and support women in "demanding their rights for dignity and citizenship."[32] MMTA-CC's activities led to the establishment of a specialized police department for women's services, a place where women could file claims of domestic violence. The MMTA-CC was also deeply involved in supporting the victims of a series of violent sexual acts committed against young boys between 1989 and 1993. The case became a political cause around which the Catholic Church and MMTA-CC organized support and fought for justice.[33]

In conjunction with the Catholic Church and new social movement organizations such as the MMTA-CC, the Workers' Party became a crucial uniting force in Altamira. The party formed in 1980, during the early stages of Brazil's transition to democracy. Around Brazil, much of the activism in the 1970s was rooted in Marxism and liberation theology and focused on grassroots organizing with a shared goal of ending poverty. The Workers' Party emerged in this context as a political vehicle through which to consolidate various struggles for rights, justice, and equality, and the party gained broad support from labor and social movements, artists, academics, and others on the left. With a commitment to representation and broad-scale participation in decision-making processes, the party became a voice through which activists' demands and movements' ideologies could be translated to the political realm. Workers' unions, social movements, and people working outside the traditional political system saw the party as an opportunity for real change to the corruption, cronyism, and clientelism that threatened the democracy they had fought to achieve. Throughout the 1980s, the Workers' Party strengthened, winning more gubernatorial races, mayoral positions, and state and federal deputy positions in each election. By 1989, when Brazil held the first direct presidential election in three decades after military rule, party founder Luiz Inácio Lula da Silva made it to the runoff stage of the election. He fell short of winning the presidency until more than ten years later, but the 1989 election solidified the Workers' Party as a major political player that could draw support from around the country.[34]

Nationwide, the Catholic Church, particularly through liberation theology and the use of Christian Base Communities, played a significant role throughout the 1980s in supporting and growing the Workers' Party. The idea that the church had organized from the bottom up aligned with the Workers' Party strategy for developing support. Given the history of collective action through the church along the Transamazon, it is no surprise that the Workers' Party became important for activists in the region. In Altamira, meetings for the Workers' Party became gathering places for people who were struggling to have their political voice heard and make demands for better living and working conditions. The Workers' Party was creating opportunities for people to meet one another, providing an avenue through which concerned citizens could get involved to better their communities, and organizing people to support the causes of the marginalized.

Irá was one of the many activists who became politically engaged through the Workers' Party. She was born in the northeastern state of Bahia and moved to the southeast of the country as a young girl. Her family made their way to Pará when Irá was an older teenager, and soon after, around 1990, she was attending meetings of the Workers' Party. As a proud dark-skinned woman who cared deeply about social justice, Irá helped create the Black Movement and was a central figure in the women's movement. Alongside Antonia Melo, she was also an active organizer of the local anti-dam movement, Xingu Vivo, during Belo Monte's construction. In my experience, she spoke directly to everybody, freely sharing her pointed critiques and dissatisfaction with the dam and the problems of Brazil. Although she came across as stern when discussing her convictions, Irá also often joked with colleagues and visitors, happy to bring moments of levity to situations she otherwise viewed as dire. She spent much of her time in the office of Xingu Vivo and so was accustomed to speaking with foreigners and visitors from elsewhere in Brazil. Irá was not as well known by the public as was Antonia and she stayed away from journalists' cameras, but she was responsible for many of the daily functions of Xingu Vivo and organized meetings and events. Irá's strong opinions and fierce attitude about and toward the political system were rooted in the Workers' Party. She described how she first became an activist in Altamira: "In 1989 or 1990, I met Mika [the woman who started the education union]. She was the first person I met who was involved in movement-related things. I was working as a teacher in a school [in Altamira], as was Mika, and wasn't a part of the movement at the time, which seemed split apart. She brought me to Workers' Party meetings, a space where everybody met together. We were discussing the environment, women, the Black Movement, and other similar issues. I started learning about all of these issues."

Irá's comments highlight the importance of the Workers' Party in consolidating the social movements in the region. Their meetings served as opportunities for engaged residents and budding activists with different priorities to speak with each other and organize events together. This coming together encouraged the emergence of local social movements to focus on specific regional issues. The growing group of activists that had roots in the Catholic Church and rural workers' union soon became a social movement of their own, the Movimento Pela Sobrevivência na Transamazônica (MPST—Movement for the Survival on the Transamazon). In 1992, the leaders, including Antonia, Simone, Mika, and others, formed Fundação Viver, Produzir e Preservar (FVPP—Living, Producing, and Preserving Foundation), providing the movement with legal rights and representation. Alongside MPST, FVPP quickly became—and remained—the principal organization fighting for better conditions for small family farmers.[35]

The region's civil society leaders worked directly with women and rural families to educate people about their rights and their power in creating change. They used this outreach to organize a broad constituency, which gave them leverage when they made demands on government regarding particular problems. Mika highlighted the importance of first gaining public support for their causes: "The issues are diverse. There are questions about the environment, about economics, about houses. For all of these, going to the streets was my focus. Public policy is also needed, but my focus is the street. You have to call on residents, get residents to push to guarantee the rights for everyone. It's not just about getting the attention of the politicians regarding the issues."

With an expanding base of support, activists would identify and target specific government agencies for particular problems. They went directly to the municipal, state, and federal officials to demand that schools be built and health posts be improved or created. Each success helped build the struggle for rights. Simone, a leader of FVPP who had been active in the movement since the 1980s, underscored this approach: "We were militants but never associated with any armed movements in Latin America... Marches? Yes, of course. There were meetings with people in the communities to hear from them. We would go to the *prefeituras* [city halls] to get schools built, etc. Go to INCRA [federal land colonization and titling agency] to resolve problems. Let's do a march for this or that... a lot of public activities. This really grew, and in 1991–1992 we went straight to Brasília, the central government, with concerns. The movement was building, building."

As the movement grew throughout the 1990s, social and political networks in the region further consolidated around advocacy for basic services. For example, local activists organized to demand more reliable electricity in the city of Altamira. In 1997, an assortment of groups gathered two thousand people in the streets to demand energy in Altamira from the Tucuruí dam, which had been completed for over a decade. According to activists, electricity in Altamira had been spotty due to its reliance on gas-powered generators, and the energy from Tucuruí was not benefiting the region that most felt the negative impacts of its construction. Activists who participated argued that it was a significant political fight, and they credited their protest activities as the main reason for the construction of the energy line that has, since 1999, delivered energy to the city.

The direct acts of protest and new activist organizations shaped organized civil society in and around Altamira during Brazil's transition to democracy, when the state continued to have minimal presence in the region. As democracy in Brazil solidified and Eletrobrás and federal agencies renewed viability studies for damming the river, these groups began advocating *for* the kinds of development along the Transamazon that would advance justice, grow the economy, and preserve the environment. In a purposeful signal of these forward-looking goals, the Movement for Survival on the Transamazon changed their name to the Movement for the Development of the Transamazon and Xingu (MDTX—Movimento Pelo Desenvolvimento da Transamazônica e Xingu). The shift from "survival" to "development" was meaningful. They were signaling that residents wanted to be active participants in planning the future of the region, rather than passive recipients of dam projects and other national initiatives they opposed. Social movement activists wanted the region to be developed, but not in a way they believed would continue to benefit outside actors at the expense of local residents. To highlight their point, MDTX organized their own research to criticize the viability studies for damming the Xingu that the federal government and energy companies were carrying out.[36]

As civil society coalesced around new visions of development and through strong social and political organizations, activists were well positioned to mount opposition to renewed state-led development initiatives in Amazonia. Regional social justice and anti-dam activists rallied together to further advocate for their own vision of development and against that of large-scale infrastructure projects after President Cardoso unveiled projects in the latter part of the 1990s that echoed the military's frontier expansion of the 1960s and 1970s. In 2002, just a few years later and during President Cardoso's last year in office, Eletrobrás and its subsidiary Eletronorte released the new plans to dam the Xingu River. Despite the new design's significant reduction in projected negative consequences, opponents still saw many risks, and dam resistance continued to strengthen. The consolidated social movement held local, national, and international events, with one of MDTX's first rallies drawing together two thousand people to protest the dam. They produced books, articles, and reports by social movement activists and academics to counter claims made by proponents of the project.[37] They held meetings to discuss the impacts Belo Monte would have on the local population and natural environment, garnered support from international organizations, including Amazon Watch and International Rivers. Some activists were confident that they could once again stop Belo Monte from being built. Construction was far from certain, as lengthy political, legal, funding, and licensing processes needed to play out, and the resistance was well organized. Furthermore, the Workers' Party was growing in strength and they could mount effective political challenges to the dam.

When Lula won the presidency in 2002, Antonia, Irá, Mika, Simone, and other dam opponents partied on the river's edge to celebrate the Workers' Party success, commemorating a political victory that had seemed inconceivable just a few years earlier. Lula's election was a victory for activists and progressive leaders along the Transamazon, as it was for leftists around Brazil, who had fought for social justice and basic needs for three decades. Lula had come to power with the support of social movements, the Catholic Church, labor unions, intellectuals and artists, civil rights activists, and many of the most marginalized and economically disadvantaged communities throughout the country.[38] In Altamira, many of these activists assumed that Lula's presidency meant that the new administration would stop construction of Belo Monte and other large dams. They also hoped that it would be easier to create alternative forms of development to support the residents in the region. The reality of the subsequent fourteen years under a government led by the Workers' Party turned out to be far more complex.

3 · DEMOCRATIC DEVELOPMENTALISM

In June 2010, just twelve months before construction of Belo Monte began, President Lula visited Altamira in a watershed moment that highlighted and deepened the rifts between various factions of civil society that had started to form in the previous few years. During the visit, the president spoke at the city's soccer stadium, the largest public gathering space. On stage, with Indigenous representatives at his side, Lula delivered an impassioned speech in which he addressed Belo Monte's critics, arguing that Belo Monte would be built in a new way that would benefit the people living in the region:

> I know that what many well-intentional people ... don't want is to repeat the errors committed in this country throughout hydroelectric dam construction. We never again will want a hydroelectric dam that commits the crime of insanity that was Balbina, in the state of Amazonas. We don't want to repeat Tucuruí. We want to do something new. So, let me give some advice to the comrades who are against [the dam]: instead of being against it, propose alternatives to use the four billion reais that we are making available in the process to take care of the social and environmental question. Let's discuss how it is that we will use this four billion, to better the life of the riverine people, to better the life of the Indians, to better the life of farmers. There are four billion reais, money that Pará has never seen, to take care of social questions.[1]

Many of the "comrades" he spoke of were not, however, allowed into the event. Antonia Melo da Silva, the leader of the Xingu Vivo movement, was one of them. According to her, former members of the anti-dam movement stood at the entrances to the stadium, donning buttons with "some sort of government logo" and serving as gatekeepers for the event. They stopped Xingu Vivo members from entering, presumably because organizers were worried that protests would disrupt the event. Antonia did not seem bothered that she was refused entry but was annoyed that former allies were the ones to turn her away. She exclaimed, "They had said they were against the dam, but . . ." She trailed off, clearly frustrated by the memory as she unsuccessfully tried to come up with a reason that they had turned their backs on the movement.

A representative of Instituto Socioambiental (ISA—Socio-environmental Institute), a major nongovernmental organization that supports traditional communities around Brazil and which has an office in Altamira, further explained the impacts of President Lula's visit to Altamira: "Everybody knew the president was going to speak in favor of the dam, so it became a confusing time for people." Some decided to stay home, others went to the stadium for the event and even helped with the logistics, and yet others protested in the streets. "It was quite apparent to people who was on what side," the ISA representative said. While he was hesitant to paint it as a black-and-white moment of people going to one side or the other, he suggested that many people saw the situation that way. As a result of this perception, a significant split occurred among resistance efforts.

In the months and years following President Lula's visit to Altamira, relationships dissolved between people who had worked alongside one another for many years. Antonia and Simone, for example, who had developed their activism through the same institutions and who had formed a movement together, no longer collaborated. Previously strong partnerships fractured to such a degree that people stopped speaking with one another and publicly rebuked former friends and allies. These splits upended claims-making efforts just as dam construction began and when the impacts were most acute, arguably the most important time to make concerted and well-organized demands.

The divisions between former allies were, in part, due to the contradictory ways the Workers' Party was governing, which Lula's speech highlighted. On the one hand, the government implemented progressive ideals, such as income redistribution, environmental protections, and initiatives aimed at including more citizens in consequential decision-making processes. On the other hand, the aggressive economic growth strategy did not seem to break from previous administrations. Instead, the Workers' Party intensified the state-led industrialization and infrastructure development approach that began in the first half of the twentieth century and that increased during the military dictatorship. This meant constructing large-scale infrastructure projects such as hydroelectric dams. The government aimed to resolve these seeming contradictions by planning and implementing projects in ways that would be markedly different from those in the past. They would institute social and environmental protections for the people and nature impacted by such projects and incorporate local voices in making decisions about future development for the region.

This democratic developmental approach to dam construction, which I describe in detail in chapter 1, came at a time when the energy sector had been semi-privatized and when the courts and licensing processes played a more substantial role in infrastructure projects. Both of these factors would have consequential impacts on why and how dam construction would continue, despite significant resistance and a fairly robust legal system that was in place purportedly to mitigate the social and environmental impacts. As a result, the administration's new approach to building dams and other projects paradoxically complicated how people made demands, upending social and political networks. More radical visions of democracy and development were marginalized, as a growing number of people worked with formal structures of engagement, and broader portions of the population could be seen as consenting to the dam's construction. The stories of the splintering of previously strong relationships and the ways the government and private interests justified and built support for dam construction provide a window into the reconfiguration of political conflict in Brazil during and after Lula's presidency.

LEFTIST ACTIVISTS COMING TO TERMS WITH A LEFTIST ADMINISTRATION

The responses to Lula's visit to Altamira highlight that dam opponents came to terms with the contradictions of the Workers' Party in conflicting ways. For decades, people had come together to protest against the dam. They had rallied in the streets and pressured the government and international actors to stop construction. They had mobilized support for the Workers' Party and celebrated President Lula's victory less than eight years previously. They had thought that, with the Workers' Party in power, the government would listen to and address their concerns regarding Belo Monte and other injustices. Now, the president that these activists had fought to elect was instituting a project they had long opposed.

The splits in civil society that were on display when President Lula spoke in Altamira had begun a few years earlier, when it was clear that the government would actively push Belo Monte's construction. At the national level, President Lula's approach to the economy during his first term, 2003–2007, was pragmatic and cautious, foreshadowing some of his decisions around development and dam construction. He supported policies that kept inflation rates low, strengthened international relations to open up new markets for Brazilian products, and did not attempt to reverse previously privatized industries.[2]

Lula also reinvigorated government intervention in industrial growth, strengthening the developmental strategy that had begun during the military dictatorship.[3] In 2007, just after Lula began his second term as president, his administration launched the Programa de Aceleração do Crescimento (PAC—Growth Acceleration Program), his aggressive development initiative, and the government started pouring resources into dam projects. By 2008, Eletrobrás, the state-led electric utilities company, completed studies on Belo Monte's environmental impacts. Large private companies and investment firms were keen to be involved with the profit-making opportunities of large-scale infrastructure. Additionally,

Brazil's Banco Nacional de Desenvolvimento Econômico e Social (BNDES—National Bank for Economic and Social Development) expanded lending for industry and infrastructure projects under Lula.[4] BNDES financed much of Belo Monte, and this internal investment in the project made it more secure. Construction was no longer reliant on foreign sources of financing, which had been a key factor in halting the project two decades previously, when the World Bank pulled its funding.

These indicators that the federal government was backing Belo Monte came on the heels of growing local support of the dam. In 2006, a coalition of business leaders in and around Altamira organized a rally in support of the dam. The march drew two thousand people, who argued that regional economic opportunities would accompany construction. Shortly thereafter, in 2007, the Workers' Party secured the governorship in the state of Pará, strengthening the political support for the project in the region.

The growing likelihood of Belo Monte's construction forced individual activists and organizations to come to terms with the political party that had once united them. With ever fewer barriers to construction, the anti-dam activists had to make important decisions: would they continue protesting against the dam or would they engage the government in efforts to mitigate the negative effects? Their decisions along these lines would have significant impacts on their relations with other dam opponents, realigning social and political networks along the Transamazon.

Some anti-dam activists, such as Antonia Melo and the Catholic bishop Dom Erwin Kräutler, remained staunchly opposed to the dam and grew to disavow the Workers' Party. They continued to generate support for protests and other anti-dam events, they refused any cooperation or discussions with the government, and they denounced the Belo Monte project and the Workers' Party altogether. In 2013, Antonia explained why she left the party four years earlier:

> The parties on the left and Lula are supposed to be of the people... They are said to be popular parties. As [Lula] came into power, it weakened the fight against the government as it changed to support the government... I won't affiliate with a party anymore. I believe in the popular organization of the people. I believe in going to the streets [*luta da rua*], fighting for people's rights, for quality education, for quality health. For the rest of my life I will fight for these things. As it says in our constitution, we have these rights... Any party that gets into power will try to keep their power. They will dominate. They want the power of domination.

Antonia was not the only one who was greatly disappointed with Lula's government. In a speech about Belo Monte, a lawyer who often works with Xingu Vivo expressed similar disappointments with the government: "I was overjoyed in 2002 when Lula was elected, and I have never felt so deceived and let down by my government." As Antonia and others like her searched for ways to reconcile their past hopes with the Workers' Party and the disenchantment they felt as Belo Monte became a reality, they refused to participate in the Workers' Party project.

Many people and organizations that had been closely aligned with these groups, however, made different choices. Instead of refusing to engage with government officials, they made the choice to remain aligned with the Workers' Party. In efforts to support the populations they represented, they hoped they could access the many resources that would accompany the dam. For example, Fundação Viver, Produzir, e Preserver (FVPP—Living, Producing, and Preserving Foundation) decided to modify their anti-dam position in 2008. About a year after dam construction began, Simone, who by that time had become a leader of FVPP, explained their decision: "We were always part of the fight against the dam. But with the building of the dam, the movement has a different posture. We have to ask ourselves questions. What's our focus? ... What will happen to the city? The families? It's a fight for the rights of the people. [Instead of staying opposed to the dam,] the other side is a discussion with the government." Simone commented on the difficult and conscious position that FVPP took in opening themselves to direct conversations with the Workers' Party government. While she commended the people and groups who continued to protest against the dam, she argued that FVPP's new conciliatory position was more difficult—and thus more worthwhile—than continued adamant opposition.

Claudio, another leader of FVPP who remained a member of the Workers' Party, was also at the center of the divisions of previously strong networks. Claudio was the son of small farmers who had migrated from the northeast of Brazil in 1958 to Santarém, a city five hundred kilometers to the west of Altamira. In the 1990s, he moved to the Transamazon near Altamira, where he became a teacher and then helped lead the MDTX, the Movement for the Development of the Transamazon and Xingu. For the subsequent decade, he was involved in social movement activity, the fight against the dam, and rural workers' unions. In 2010, he took on a leadership position in FVPP. He shared perspectives similar to those of Simone, explaining, "We [the people fighting against the dam] were beaten because our biggest goal [of stopping the dam's construction] did not happen. The political force of the government was bigger than our force of the people ... We are small in comparison." But he stressed that it is important to recognize that the fifteen years of civil society work against Belo Monte was not done in vain. He listed off the changes in plans that reduced flooding, the list of conditions that Norte Energia had to meet in order to receive the license to operate the facility, and the increased opportunities for participation with government officials.

Despite Claudio's, Simone's, and Antonia's shared belief in the power of popular organizing, their mutual desire to bring unity to their struggles for the people in the region, and their long history of working together, the connections between their organizations severed as dam construction approached and each dealt with the changes in the Workers' Party in different ways. I often heard Xingu Vivo express their disappointment in those who, like Claudio, had remained loyal to the Workers' Party and who were engaging in any kind of direct participation and negotiation with government actors or Norte Energia.

The ways these local individuals and groups came to terms with the characteristics of the contemporary Workers' Party government also impacted how they worked with groups from outside Altamira. This was particularly true of the relations between Xingu Vivo and the national group Movimento dos Atingidos por Barragens (MAB—Movement of People Affected by Dams). MAB has opposed large-scale infrastructure development and hydroelectric dams since their founding in 1991. The movement has influenced national-level policy and organized dozens of communities in anti-dam movements, even preventing some dams from being built. MAB also works with communities who will be affected by dams in order to support and educate residents as they make demands on their local government. Despite their opposition to dams, MAB continued to support the Workers' Party during Lula's presidency, even as the government aggressively pushed large-scale development projects. In 2010, the movement backed the party's presidential candidate, Dilma Rousseff, in spite of Rousseff's history of expanding the energy sector's free market and advocating for large scale power generation and distribution projects when she was minister of energy and Lula's chief of staff.[5]

For local anti-dam activists, these positions equated with support, or at least acceptance, of Belo Monte.[6] The role of MAB in and around Altamira was further complicated because of the movement's relatively late arrival to Altamira. MAB had spoken out against Belo Monte from afar but did not send a representative to Altamira until 2009. Once there, the movement's leaders began organizing urban residents who faced displacement. Rather than advocate against the dam, however, MAB spent their energy creating a public base of support. Their practice of using "base communities" was designed to bring local residents together to support each other, identify and empower leaders, and build a groundswell of support for making demands about pressing issues. Movement leaders held organizational trainings for local leaders to encourage and train the community to leverage local politicians and recruit new residents to join the movement. One of the longtime national MAB organizers who arrived in Altamira in 2012 explained MAB's work with the communities, despite the deep philosophical problems the organization has with the dam construction:

> Brazil is governed by capital. Decisions are made in favor of capital, so the issues go well beyond Belo Monte. And the issues here are more than just the actual dam. It's about energy, the actual dam, and transmission of the energy . . . The problem isn't necessarily the particular government but the government has a responsibility

to better conditions for the people ... MAB is against the dam because it is an environmental and social crime, but they fight for the people to live a good life, so they don't just stop at being against the dam. The problem is the model being used by the government, a model based on capital. Capital is the problem.

One would think Antonia and the others at Xingu Vivo would agree with MAB's perspective. After all, they used the same language of Belo Monte as a "crime" and are both fighting for the people affected by the dam. However, while they occasionally worked together, they did not see eye to eye on many things. MAB's approach to supporting the local population strained potential relations with other social movement organizers, in part because of MAB's continued affiliation with the Workers' Party. These reasons became clearer in my interview with Antonia. "MAB says they are fighting big capital and that to do so the country needs to move toward a more socialist society," Antonia said. "But here, in these cases, big capital has names and addresses. There are names of big capital: Lula, Dilma, the courts, and the company, Norte Energia. MAB defends the Workers' Party—they say it's not the fault of Dilma and Lula but that it's just big capital. But it's not. I've been saying this for a long time."

Xingu Vivo wanted nothing to do with the government or the Workers' Party. This caused significant tension with MAB, who found it possible to engage in both contentious activities and more direct participatory processes with government officials. MAB and FVPP leaders told me that labeling people as either in favor of or against the dam disrupted the ability to organize because it created enemies and made it difficult to have a more complex view. They often stressed that their groups are not political, in that they are not officially associated with a particular political party, but that activists within them would share personal opinions. One such MAB organizer explained, in the most diplomatic way possible, "The Workers' Party was born of social movements but the Workers' Party program has since been abandoned. That said, I think Workers' Party is the best option for now."

The developing divides between civil society groups were not only based on ideological positions related to the Workers' Party. The fissures were also a result of the differing responses to the government's promises of new resources and expanded opportunities to participate in meaningful decision-making processes. President Lula's speech addressed these opportunities directly, stressing that Belo Monte would be different from past dams because it included an unprecedented amount of investment in mitigation and regional development efforts. He also indicated that the government wanted to hear from the people. In referencing these financial and participatory opportunities, Lula was directly confronting the anti-dam protesters, a moment that would foreshadow the growing divisions within resistance efforts.

President Lula's rhetoric of opportunity resonated with, and was repeated by, many people involved in the conflict over Belo Monte and local residents who would be most affected by construction. For decades, the Transamazon area had been largely abandoned by the government, leaving residents to fend for themselves. Many local residents, even those opposed to the dam, felt as though Belo Monte could finally bring needed, even if temporary, attention that could be used to advance the well-being of hundreds of thousands of people. In discussing FVPP's decision to adjust their position on the dam, Simone once commented that "every big project has resources for mitigation of the effects." She suggested that once dam construction was imminent and underway, it was best, and possibly even more challenging and potentially rewarding, to harness those resources, rather than remain adamantly opposed to a project. "It's more comfortable to stay against [the dam], to stay the same line and continue being against Belo Monte," she said. "That is a really strong and brave position, but comfortable, too. But thinking about people ten to fifteen years from now, that's not easy. And it's not just here. The whole region will see big changes." Simone and others felt it was important to do the difficult work of deciding the kind of future that the community wanted and how activists could help bring that about.

For staunch dam opponents, however, the government concessions were simply window dressing, meant to obscure the damage that would be caused by Belo Monte and distract people just long enough to build the dam with less resistance. These steadfast opponents were highly critical of those who were deciding to take any resources from the government, arguing that the attention and resources would be short-lived and fail to bring any meaningful change. With construction approaching, the rhetoric of resource abundance was a critical point around which political conflict was being reconfigured.

UPHEAVAL AND ENGAGEMENT

Once construction began, the divisions within civil society only intensified. Resources began flowing into the region, new organizations emerged, outsiders from a wide range of places arrived, the implementation of mitigation efforts started, and participatory initiatives began. The changes to the area, while predictable, were stunning and fast paced. I immediately recognized some of the transformation from the moment my plane landed in July 2011, less than a year since my previous trip to Altamira and a month after construction commenced. The small airport was full of people who were clearly not from the region, evidenced by the badges they were hanging around their necks that displayed the names of engineering companies that had subcontracted aspects of dam construction work. The taxi from the airport to the city center had doubled in price in less than a year, as had the price of the hotel. The cab driver explained that

I was lucky to have reserved lodging in advance, as there were no vacancies in the city. The streets, which rarely had much traffic in the past, were filled with buses, motorcycles, cars, and new pick-up trucks, many of which had license plates from other states. Every corner seemed to have a restaurant or hotel under renovations and another in the early stages of construction. The city was teeming with energy.

The beginning of dam construction had also brought intense yet disparate emotions and perspectives to the people who lived there. On the one hand, there was a great deal of excitement and hope. People were arriving from all parts of the country, new businesses were opening, and people were looking forward to new schools, health care facilities, and other basic infrastructure projects that the government promised would accompany dam construction. Many local residents were relieved that the state and federal government were finally paying attention to the needs of local people. On the other hand, the changes to the region were disruptive. Given the population explosion, prices for rent and everyday goods were skyrocketing, raising the cost of living. The apartment I found to rent, for example, was over three times the cost it would have been just a year previously. Many food staples doubled in price. Buses had begun transporting workers from the city to the work sites, kicking up and covering the city in red dust. People moved from nearby towns to find work on the dam, leading to labor shortages elsewhere.[7] Traffic accidents were increasing in frequency and severity; every night the news showed mangled motorcycles and injured people. Violence was on the rise, and people worried about an increased flow of drugs through the area. Close to forty thousand families would be displaced in the subsequent few years.

The social and political landscape was changing as well. New people seemed to be arriving every day, and both government and citizen groups were beginning new initiatives to advance their goals, ensure that the dam would bring positive benefits, and reduce negative impacts. Street rallies and protests seemed to be happening every few days, even as local social movement activists were in conflict with one another. Local, national, and international activist groups were attempting to find the ways they might best be able to support their agendas. The federal prosecutors' office was putting together new claims about Belo Monte in an effort to stop construction. Municipal government officials were struggling to convince the population that the dam project could bring positive change while they also complained about the lack of federal investment in the area that could have prepared the region for these changes. Federal officials were arriving from the south of the country, some of whom argued that the government was finally offering necessary services to the local populations while others met with local people to see how they could best serve them. Officials representing the construction consortium in charge of Belo Monte, many of whom had just moved to the city, were touting the project as an opportunity for the region and arguing

FIGURE 3.1. Protestors at the Norte Energia headquarters in July 2011. Large sign reads, "Women of the Transamazon and Xingu in the Fight for Life against the Belo Monte Dams." Smaller flags say, "No to the Belo Monte Dams on the Xingu." (Photo by author.)

that they would be supporting the local population and the country as a whole. Meanwhile, everyday people were attempting to determine how their lives would be affected, hoping that the rhetoric of local development and opportunity would actually come to pass and worrying that they would instead be faced with an even more tenuous way of life.

Amidst the sudden upheaval that accompanied dam construction, the first meetings of the Comitê Gestor do Plano de Desenvolvimento Regional Sustentável do Xingu (CGDEX—Steering Committee of the Regional Sustainable Development Plan of the Xingu) took place in Altamira. This led to a further splintering of civil society. The federal government had designed the committee as the participatory initiative to guide regional development. Tasked with allocating five hundred million Brazilian reals that Norte Energia was mandated to provide, CGDEX offered significant financial resources and a chance for local groups to influence and create projects that would impact the region. The committee also offered an opportunity for local groups to engage directly with government officials. The thirty-member voting body was made up of representatives from regional civil society organizations and government officials from municipal, state, and federal levels, and other members of the community could attend, participate in the monthly meetings, and partner with official members to propose projects. Given the amount of financial resources involved and the unique opportunities for engagement with federal officials, local organizations and individuals responded to CGDEX in varied ways. In turn, divisions deepened and new alliances formed.

CGDEX was appealing for many organizations and dam-affected residents who had historically felt abandoned by the state or those who had advocated for the dam but were worried about the negative impacts. Many people and organizations jumped at the chance to participate, hoping to secure resources for their organization or contribute to decisions about how to develop the region. Many of those people had been quite critical of both the local and federal governments, but viewed CGDEX as a new opportunity to create meaningful development and for marginalized populations to have their voices heard. Jonas was one such civically active member from the region who was enthusiastic about the possibilities of CGDEX. As a businessman, entrepreneur, and a leader of FORT Xingu, a group composed of over 130 businesses and industry leaders, his positive feelings about CGDEX were not surprising. It *was* surprising that he was actually quite critical of the dam and worried about the consequences of construction. He told me that he had visited fourteen other locations where dams had been built and all of them had the same story: "The dam comes, money comes, people come. There is a rise and then just after the peak there is a 20 percent decrease in the value of goods and then 30 percent of businesses break." He talked about the unemployment problems after the height of construction and the negative impacts in the long term. CGDEX, he explained, was the only thing that could bring different outcomes from previous hydroelectric projects. Jonas believed that the region would see an overall negative impact if people from the region did not participate in the process.

Others were somewhat more reluctant participants initially but grew to fully embrace CGDEX as a crucial opportunity for development in the region. Claudio, the director of FVPP, explained that in 2010, CGDEX "was still just a presidential decree. We had to transform it into a reality." He said that CGDEX became a place where "civil society can arrive, speak, share opinions, present projects, and get resources to develop and strengthen society." Claudio took a key leadership role of CGDEX as a representative of the civil society group.

Instituto Socioambiental (ISA), the nongovernmental organization that works to support traditional communities, was another civil society organization that unenthusiastically agreed to participate in CGDEX. In 2011, just as CGDEX was about to hold its first meeting, one of ISA's founders told me that although they were accepting the invitation to participate in CGDEX, they were prepared to withdraw if they did not feel right about what was happening in what was then a new participatory space. They had sensed that the government might be using CGDEX merely as a way to appear as though they were taking the public into consideration, rather than actually doing so. Despite the concerns, ISA wanted to take the opportunity to access the decision-making processes and what the founder called the "inside." Rather than using the forum for the potential financial and material resources it could offer, however, their focus was on bridging government and civil society. By framing it this way, ISA was trying to assuage concerns that civil society organizations were only participating in CGDEX to access financial resources. Instead, they focused on the opportunities to engage in meaningful conversations about the future of the region. The group maintained this position, and after a year of participating, one of the representatives told me that "CGDEX is one of the only places where people are actually mobilizing. It's a place of debate and dispute." They remained cautiously optimistic and maintained their position that CGDEX could at least offer opportunities for diverse people and groups in the region to come together.

Not surprisingly, the staunch anti-dam groups like Xingu Vivo did not see CGDEX as an opportunity. Rather, for them it was a ruse to further stifle resistance and divide civil society. On a few occasions, anti-dam activists used CGDEX as a target of protest, but they never joined the dialogue or worked with the technical committees that did much of the work of CGDEX. After one such protest of the CGDEX meetings, Antonia Melo told me about a brief conversation she had with the federal official who organized and facilitated CGDEX meetings: "He says that I don't want dialogue. The dialogue the government wants isn't really dialogue. It's just a way to get what they want!"

Furthermore, these oppositional groups loathed the idea that official civil society participants of CGDEX represented the community and dam-affected groups as a whole. For example, these activists publicly decried participatory processes during a community forum about urban displacement in April of 2012. The forum had been called by Jonas, a leader of FORT, and he invited Norte Energia representatives to give the community details on the resettlement plans. Jonas stressed to an audience of over a hundred residents, activists, and leaders of organizations that the participatory processes associated with Belo Monte were creating stronger relationships between the various stakeholders and making the project more successful with fewer negative consequences. I sat next to a few activists, including Irá, the active member of Xingu Vivo and leader of the local Movement of Black Women. As Jonas made his comments about participatory activities, Irá became visibly frustrated, shaking her head and joining others in yelling "Lies, Lies!" and complaining that this was not a democratic process. Claudio of FVPP later commented during the forum about CGDEX and how it was crucial for civil society leaders and residents to continue participating. Irá was so disgusted at this point that she left the forum altogether, remarking the next day that she was frustrated that FORT and FVPP pretended to be representing the community. CGDEX and the idea that the government and Norte Energia had any interest in meaningful dialogue that would help the community seemed insulting to her. Given FVPP's leadership role in CGDEX, this further deepened the divides between the groups.

A few important actors in the claims-making processes around Belo Monte, such as the public defender and public prosecutors, remained completely outside of CGDEX. These state-funded yet autonomous legal agencies can bring cases against the state or private actors to support vulnerable communities or to enforce regulations. In order to do so effectively, they strive to remain as independent as possible. "We don't want to be associated with CGDEX," Andreia Macedo Barreto, the public defender from the state of Pará, explained. "The money [for CGDEX] comes from the company, so participating would make it harder to do our work. Whether or not to participate was a big discussion [within the public defenders' office] and we talked about it a lot, but we need to stay independent." Andreia and the other public defenders felt as though they should remain as neutral as possible. Given that the public defender often brought cases against Norte Energia, any kind of participation could be problematic, particularly because of the funding stream for CGDEX.

In sum, CGDEX represented a new type of collective governance. It allowed various if limited segments of the population to direct how financial resources were allocated during a moment of significant change. CGDEX was also emblematic of the democratic developmental state that aimed to combine inclusive decision-making practices while pushing forward large-scale infrastructure projects. As one of the few examples of a state-created opportunity to give some decision-making power to local residents in working toward sustainable development, CGDEX can be seen as the most recent step in the management of resources in and development of the Xingu region. On the one hand, CGDEX offered an institutionalized form of engagement that empowered local residents to have a say in their own future, a step forward in a region that has a long history of outsiders' extraction and exploitation. On the other hand, CGDEX also served as a way for the state to co-opt dam resistance and the claims-making efforts of civil society and social movements. As a result, in addition to becoming a primary site of negotiation between residents, activists, local organizations, and government officials, CGDEX realigned, and also represented the realignment of, social, political, and civic networks in the region. The choices that each group made to participate or remain outside, and their subsequent opinions about CGDEX, deepened the divides that had been forming during the previous few years. In this way, the existence of CGDEX was another factor that upended collective mobilization and claims-making efforts, fracturing previously strong networks but also creating new and surprising alliances.

PRIVATIZATION AND DISPLACED RESPONSIBILITY

Development efforts in and around Altamira were also complicated by the ways responsibility for mitigation and adaptation programs was shared between the Norte Energia consortium and the various levels of government. As I drove through Altamira and across the countryside during the first two years of construction, I noticed many newly built school buildings. Signs hung outside with the logo for Norte Energia and a description of the program that led to the school's construction. While the number of new schools in the region was significant, they were not surprising to see. The consortium was required to build them as part of their contract to construct the dam. Most of these schools, however, remained empty or unfinished. Rarely, if ever, did I see children in those schools. Norte Energia representatives told me that the construction consortium was only responsible for building the physical structures. The municipality was responsible for staffing and maintaining the schools, but local government officials claimed they did not have the funds to operate them. Similarly, Norte Energia built a city sewer system for Altamira, something residents had demanded for decades, but the city was responsible for operating it. Furthermore, Norte Energia would not pay for all of the connections from the houses to the city system, an expensive part of the system.

As these examples demonstrate, neither Norte Energia nor municipal governments took responsibility for completing the projects that the federal government and the consortium had promised would accompany Belo Monte's construction. Norte Energia was operating from the expected ideology of a large company. While they were required to provide certain mitigation and development efforts, they wanted to spend as little money as possible and thus relinquish responsibility to the municipal, state, and federal governments. Norte Energia cited the limitations of their projects, suggesting that the local governments were responsible for the welfare of their communities. Conversely, the municipal governments were often unwilling, or claimed they were unable, to oversee or carry out projects associated with Belo Monte. At times, local politicians argued that projects related to Belo Monte were the responsibility of Norte Energia, citing the unprecedented resources that the construction consortium possessed. At other times, the municipal government cited a lack of resources to complete projects. The municipality may, in fact, have lacked the funds. As laid out in the 1988 Brazilian constitution and subsequent laws and guidelines, some of the financial compensation promised to local governments would come through monthly payments over two decades, rather than during construction.[8] Additionally, the Belo Monte facility is located in the neighboring municipality of Vitória do Xingu, even though the urban center of Altamira was arguably more directly impacted. As a result, a significant portion of Norte Energia's financial compensation would go to the smaller and less populated municipality.

Due to these structural conditions, the government was able to cede responsibility by claiming weakness. The manner in which the state relinquished responsibility can be viewed as an example of what Shalini Randeria calls the "cunning state." Rather than a weak government that has no power to enforce or support the people, the cunning state is one that "seeks to redistribute responsibility" by *claiming* to be weak and unable to meet the needs of vulnerable citizens. The cunning state uses *perceived* weakness in order to remain unaccountable.[9]

These dynamics were, in part, a result of the privatization efforts of President Fernando Collor de Mello in the early 1990s and President Fernando Henrique

Cardoso later that decade. Their endeavors led to the partial privatization of the electricity industry, which set the stage for public-private partnerships in hydroelectric projects.[10] As a result, Brazil's twenty-first-century dams have been driven by the government led by the Workers' Party, but they have not been completely government run. Each has been constructed and operated by a consortium of public and private companies brought together to oversee that particular facility. In nearly all cases, private companies have held more than 50 percent of the stake in the consortium.[11] Like much of the semi-privatized energy sector, Norte Energia is a consortium of public and private entities created for the purpose of constructing the Belo Monte facility. The private companies held just over 50 percent of the total project.

On the surface, the semi-privatization of projects like this might appear not to matter all that much. The state has ultimately maintained control of the energy sector, and corruption that has been endemic to these projects has only deepened the sense that the state maintains its influence. The Lava Jato criminal investigation, which began in 2014 and disbanded in 2021, exposed how government officials took bribes from, an in turn awarded contracts to, some of the largest construction companies in Brazil, including those involved with Belo Monte.[12] Additionally, the private sector has always been a part of what Kathryn Hochstetler refers to as the "energy-building enabling coalition," a mix of state and market actors whose shared goal is to build energy projects.[13] Indeed, throughout the course of my research, government officials, company consultants, and other people familiar with the financing and management of Belo Monte told me that, regardless of the technical details, Belo Monte is a project of the federal government.

Despite the state's influence over the project, the people who were impacted by dam construction viewed Norte Energia as a private company that was separate from the state, and the technicalities that put a majority-private entity in charge of Belo Monte made claims making more challenging. The government may have advocated and guided the dam's construction, but Norte Energia and the many contracted companies that worked on the project and mitigation efforts had a great deal of control over what happened in and around Altamira. Responsibility often fell on private subcontracted companies to organize and carry out community information sessions about their projects, to document those who would be directly impacted by construction, and to complete the necessary mitigation programs. When these programs were not completed and promises of new development failed to materialize, government officials pointed to Norte Energia as the responsible party.

The state had not only a private entity to hide behind, but also CGDEX, another resource-rich avenue of regional investment created by the state, funded by Norte Energia, and driven by participants, most of whom were not government officials. As the government failed to provide the resources that the region had long desired, citizens began making demands on Norte Energia, rather than the government, and started proposing projects to CGDEX that the government should arguably undertake. This situation allowed the government to further withdraw from their responsibilities. For example, concerned activists were able to convince CGDEX during the first few meetings to create a technical committee dedicated to health concerns in the region. In the first four years of dam construction, tens of millions of CGDEX dollars were then dedicated to bettering the health clinics and hospital in and around Altamira. The local government had long claimed it was unable to provide the necessary resources to improve healthcare, so CGDEX stepped in. CGDEX did this type of work in a number of areas, and some of its members were not afraid to admit this. In 2017, for example, Claudio, who worked for FVPP and was a key representative of the civil society contingent of CGDEX members, explained, "The government should be doing some things, but since they are not, CGDEX is filling the gap." In the opinions of Claudio and others who were once reluctant to participate, the government was unlikely to carry out crucial infrastructure work, so it was important for CGDEX to do so while the resources were available. On the other hand, as CGDEX stepped in to carry out some important projects but not others, and as Norte Energia seemed to be responsible for the region's infrastructure, citizens were further confused as to how, and to whom, to make claims and demands.

This confusion is, in part, a result of what Evelina Dagnino refers to as "the perverse confluence" of the neoliberal political project and the democratic participatory project.[14] Neoliberalism is an approach to governing and a political movement that is based on the idea that the unimpeded market is better at generating economic growth and providing social welfare than the government. The state reduces its role in providing social provisions to its citizens, while private companies and individuals assume more of that burden.[15] In Brazil, neoliberal approaches to governing began with the Collor administration in 1989, expanded through the end of the Cardoso administration in 2002, and in many ways have continued ever since. At the same time, notions that democracy should be enlarged through participation have permeated Brazil. These ideas were imbued in the Brazilian constitution of 1988 and strengthened under the Workers' Party leadership. The two very distinct and often contradictory political projects obfuscate the government's responsibilities and thus complicate claims-making processes.

The ways the state and private companies displaced their responsibility helped powerful actors maintain economic and political control. Dam-affected groups and their advocates were forced to navigate a confusing situation when making claims, and the responses to their demands slowed. Not only could Norte Energia and the state argue that they were not responsible for any particular claim, they could also suggest that, due to the multiple parties involved, the process of responding to demands would take more time. In addition, CGDEX, which had emerged as a primary avenue through which to be engaged, was designed neither to make decisions quickly nor to implement projects in short fashion; rather, CGDEX took a long-term view of development and would stress a democratic process over short term gains.

The slow claims-making process, the displacement of responsibility, and the general challenge of making demands during dam construction were a result of the democratic developmental approach that the Workers' Party took in instituting Belo Monte. By offering material concessions and options to participate in decision-making processes to dam-affected populations and the region, the Workers' Party ensured that at least some individuals and groups would slow or cease their resistance in favor of new opportunities. This was successful, in part, because many of the historically ardent dam opponents maintained their allegiance to the Workers' Party, viewing the opportunities as a step toward meaningful and sustainable development in the region. Through party allegiance, the state was thus embedded in previously resistant parts of society. These mechanisms facilitated a pro-dam coalition of public and private actors—including the central state, local government, energy companies, construction companies, engineering firms, and regional elites, among others—to effectively support construction.[16]

THE LAW AND DEMOCRATIC DEVELOPMENTALISM

The democratic developmental approach to dam building was further complicated by the increase in robust laws designed to reduce social and environmental damage from large-scale infrastructure projects, as well as by the gradual strengthening of institutions to enforce those laws. The difficulties in managing the impacts of privatized projects in Brazil are not due to the lack of clear regulations, as a series of environmental and licensing laws have been created since the end of the military dictatorship. In perhaps the most significant step toward processes that would provide social and environmental protections in the context of development initiatives, the Conselho Nacional do Meio Ambiente (CONAMA—National Council on the Environment) enacted specific environmental impact assessment (EIA) regulations in 1986. This new framework for EIAs created the basic guidelines that firms must meet before state agencies will grant licenses to construct and operate projects like dams. These guidelines revised previous screening and scoping processes, mandated that independent multidisciplinary groups prepare environmental studies, and introduced a public consultation process, a novel idea at the time. The timing of the new regulations was important, as the federal constitution of 1988 embodied the social and environmental components that were integral to EIAs.[17]

Impact assessments and the associated licensing processes have remained an important part of large infrastructure projects, especially dams, and have become a core site of conflict. On one side, the coalitions that advocate and plan the construction of dams use impact assessments as a way to show their compliance with the law. On the other side, resistance movements utilize the process of assessing environmental risk, as well as the irregular implementation of these assessments as tools to make demands, change project outcomes, or call for a stop to projects altogether.[18]

The case of Belo Monte highlights the ways the licensing processes could bring about material benefits and reduce harmful impacts of infrastructure projects but could also serve to justify construction, generate consent from the population, and provide political protection for elected officials. Eletrobrás and Eletronorte completed viability studies for Belo Monte in 2002, but multiple federal courts, including the supreme court, suspended the EIA, which stalled the project. Three years passed before Brazil's National Congress, with support and insistence from President Lula, authorized Eletrobrás to complete the EIA, which they then began in 2006. In 2009, Eletrobrás published the results of the assessment in nearly twenty thousand pages across thirty-six volumes, as well as in the mandated *Relatório de Impacto Ambiental* (*RIMA—Summary of the Environmental Impact*). *RIMA*'s two hundred colorful pages filled with diagrams, bright illustrations, and maps are designed to be accessible to, and understood by, the entire population. The distribution of *RIMA* was followed by a series of public hearings, in which the Instituto Brasileiro do Meio Ambiente e dos Recursos Naturais Renováveis (Ibama—Brazilian Institute of the Environment and Renewable Natural Resources), the federal environmental agency that grants the licenses, shared information with, and purportedly listened to the concerns of, the community. Some participants and scholars called the public hearings on Belo Monte scams, stating that Ibama allowed little actual public participation. The primary hearing in Altamira was held in a room too small to accommodate affected populations, there was an overwhelming presence of police and military, and some Indigenous communities decided to leave because they did not want to be perceived as agreeing with the dam. Furthermore, engineers shared technical presentations for hours, while public comments were only allowed at the end, which was in the middle of the night. Scholars have argued that, rather than encouraging genuine public participation, these hearings actually served to legitimize the interests of the federal government, while strategically silencing opponents. In addition, these hearings provided an opportunity for state and municipal politicians—who would be up for reelection in the subsequent year—to persuade the public that the project would benefit the Xingu region.[19]

Not only were activists and scholars critical of the public consultation process, they were also skeptical of the analyses of impacts presented in the EIA and the *RIMA*. In a direct response to the official assessment, a group of researchers published a series of studies that highlighted the detrimental effects they predicted would accompany Belo Monte. The group, called the Panel of Specialists, released their report just a few weeks after the controversial public hearings.[20]

Despite the opposition, Ibama granted the preliminary license for Belo Monte in 2010, which allowed for the public auction of the project. At that time, a consortium of public and private companies created Norte Energia, which won the rights to construct and operate the dam and sell the energy that the dam would eventually generate. Over the subsequent year, the consortium developed the 2,500-page *Projeto Básico Ambiental* (*PBA—Basic Environmental Project*), which is a detailed account of the predicted impacts of dam construction and the dozens of plans to mitigate those effects.[21] The PBA allowed Norte Energia to receive the installation license in June, at which time construction commenced. Together, the preliminary and installation licenses provided the conditions that the consortium was mandated to meet and indicated the timeline by which those conditions must be met.

Officials and dam proponents often used the conditions laid out in the PBA to justify Belo Monte's construction, highlighting how the programs and projects in the PBA would benefit the region. For example, during the community forum on urban displacement, Jonas, the leader from FORT, concluded his presentation with a slide that listed the benefits of Belo Monte: basic sanitation, revitalization of the waterways, relocation of people living in at-risk areas, and better health and education. "This is a part of the PBA, so these commitments have to be met," he summarized. The murmurs of discontent from the crowd made it clear that many people were skeptical that these promises would actually be fulfilled. During the question-and-answer period, Antonia made her opinions clear: "The PBA stipulates that a few million reais will be spent on programs for the displaced population. For a project that will cost thirty billion reais, this is an embarrassment!" Antonia's comments were common among skeptical residents and activists. Not only did they think the PBA would be neglected, but they also argued that, even at its best, it would not do enough to compensate dam-affected populations and certainly would not bring the kind of development to the region that most people wanted or needed.

The impact assessment and licensing procedures were contested not only by activists, scholars, and concerned residents. Given that these procedures are built on formal laws, these processes were also disputed in the courts and on legal grounds. Most of the legal challenges came from the Ministério Público Federal (MPF—Federal Public Prosecutors' Office), a state-created and state-supported legal agency. Public prosecutors also work at the state level, such as the Ministério Público do Estado do Pará (MP-PA—Public Prosecutors' Office of the State of Pará), but the MP-PA was not as involved in Belo Monte as the MPF. The constitution of 1988 gave the MPF a breadth of functions and significant levels of autonomy from other parts of the state. Some scholars have called this legal institution the "fourth branch" of the government because one of its key functions is to keep other branches in check through judicial processes. Public prosecutors defend the interests of marginalized groups, the environment, and the general population by enforcing labor, environmental, and other regulations. They carry out their own investigations and bring civil lawsuits against individuals, companies, the police, and the government to prosecute criminal actions. They also organize public hearings to share information of public interest, including the results of research they help coordinate, and to learn more about situations that impact citizens. Lesley McAllister argues that public prosecutors "make the law matter." She points out that social and environmental laws in Brazil are quite extensive and offer a great deal of protection for marginalized groups and the environment, but these laws would be ineffective without the public prosecutors.[22] Since 2001, federal public prosecutors have brought dozens of major cases against the construction of Belo Monte, making various arguments about the illegality of the dam and the processes through which the licenses were granted to allow dam construction and operation.[23] Some of these legal challenges suspended the construction or operation licenses, leading to temporary stoppages to dam construction.

The MPF was already active in Altamira, but the office expanded during dam construction, in order to deal with the influx of issues associated with the dam. Thais Santi, the prosecutor who worked closely with the riverine people, arrived in 2012 and focused most of her attention on matters related to Belo Monte. Much of the MPF's work on dam-related matters addressed the failure of government agencies and Norte Energia to protect the environment from negative impacts that result from dam construction, the rights of Indigenous populations in the area, and other affected populations, such as the fisher and riverine communities. In 2013, for example, prosecutors brought a case against Norte Energia regarding the resettlement of people who would be displaced by the dam, citing irregularities in the construction of new homes, the absence of information, the lack of debate with those affected by construction, and the reduction of resettlement options for affected families. As with this example, prosecutors brought many cases against Norte Energia, but other MPF cases also targeted various government agencies, including Brazil's environmental protection agency (Ibama), the agency that is designed to support Indigenous rights (FUNAI), and Brazil's national development bank (BNDES).

In addition to the MPF, public defenders also emerged as key actors in the conflicts related to Belo Monte, adding another dimension to dam construction in early twentieth-century Brazil. Like public prosecutors, the public defenders' role in Brazilian society has strengthened since the late 1980s. The most recent constitution of 1988 was the first time the public defenders' role was formalized at the federal level and given the necessary resources to carry out its functions, which is to provide constitutionally granted free legal support to those in vulnerable situations and those who cannot afford it. Similar to public prosecutors, public defenders operate at the federal level, as the Defensoria Pública da União (DPU—Public Defenders' Office of the Union), and in many states, including Pará, as the

Defensoria Pública do Estado (DPE—Public Defenders' Office of the State). Public defenders can support either individuals or a group through civil suits, but unlike public prosecutors, they must be representing an individual or defined collective—rather than diffuse or collective interests—in order to bring cases to court.

The DPU and DPEs have made much slower and more scattered progress than the MPF in establishing themselves as robust institutions since states first created public defenders' offices in the 1950s. The last two states to create public defenders' offices only did so in 2011 and 2012, and multiple states did not have functioning offices for years after that.[24] The agency remains an inconsistent and mostly underdeveloped institution across the country, making the work of public defenders difficult. The state of Pará established their DPE in 1983. For two decades, public defenders in Pará were paid as little as one thousand Brazilian real (less than five hundred U.S. dollars) per month. This low salary made it difficult to attract good candidates for the job, and according to some lawyers who held these positions, it also led to high levels of corruption and bribery. In the mid-2000s, after the state significantly increased the pay and resources, the DPE began attracting better public defenders and started to become a respected and effective organization. The agency also changed its structure so that it could meet needs throughout the state, rather than only near the capital, Belém. More offices opened around Pará, and the quality of work improved.

The first DPE office in Altamira opened in March of 2011, just a few months prior to Belo Monte's ground breaking. A year later, in the midst of the rapid changes in Altamira that were accompanying the first year of dam construction, the DPE created the Grupo Especial de Trabalho Belo Monte (Special Belo Monte Working Group). This group, officially composed of nine state public defenders, was designed to create a temporary institutional space within the DPE's structure to focus on the needs of the people affected by the dam's construction.[25] Andreia Barreto, the public defender based in Altamira who supported the fishers during their monthlong protest and the conciliatory hearing in 2012, became the face of the group and the primary public defender supporting dam-affected populations. In 2015, the federal DPU finally opened an office in Altamira. Throughout the remaining years of construction, both levels of the public defenders' offices brought cases against Norte Energia to demand more just compensation and relocation processes for those impacted by Belo Monte.

Together, the public prosecutors and the public defenders working on issues related to Belo Monte, in conjunction with the legal system more broadly, further complicated dam construction and the claims-making processes of dam-affected residents. On the one hand, the Belo Monte dam has been built in the context of robust laws and legal institutions to enforce those laws. Legal processes have supported residents directly impacted by construction and have provided growing awareness of Belo Monte to the national and international community. Nearly every time construction has been halted through a legal challenge brought about by the MPF, DPE, or DPU, the media covered the story. Additionally, some of these cases have prompted Norte Energia or government agencies to change behaviors and provide some protections for the people or environment.

On the other hand, the overall legal processes have not been able to stop the dam from being constructed, nor have they prevented the dam from causing a host of negative impacts. Legal processes are slow and have rarely, in the case of Belo Monte, resulted in outcomes that favor the most marginalized. In fact, state actors and Norte Energia officials threatened activists, journalists, and residents with imprisonment and fines for nonviolent protests against the dam. At their worst, proponents of Belo Monte have used the environmental impact assessments, licensing processes, and the courts more broadly to justify construction in the name of economic growth, despite the work of the public prosecutors and public defenders.[26]

DIVISION, CONSENT, AND RESISTANCE

The ways Brazil's democratic developmental state planned and implemented the Belo Monte dam show us how the state and national elites use the politics of abundant resources and opportunities to remain powerful and further their projects, regardless of laws, opposition, and negative impacts. The rise of the Workers' Party to the presidency, the ways the new government implemented the construction of Belo Monte, the influx of resources and opportunities, and even the supposed support of the courts thus had the paradoxical effect of weakening collective mobilization and claims-making efforts. As some activists chose to remain aligned with the Workers' Party, attempted to access new resources, and engaged in innovative participatory opportunities, others wanted nothing to do with the Workers' Party government, which they saw as succumbing to power and capitalist interests. These dynamics drove a wedge in formerly united and strong networks, weakening dam opposition and claims-making efforts.

Many of the activists and engaged civil society members were well aware of the tactics the state and private companies were using to maintain control and build the dam. Claudio, of FVPP, knew the state was dividing groups, yet he chose to engage. "This question of 'who is for and who is against' is a strategy to divide society and transform the debate over this big project into a competition, as though it were a soccer game," he told me in 2012. He wanted civil society to come together and work to ensure the rights of the people were met. In order to do so, he remained an active member of the community and engaged in discussions with the government and Norte Energia, serving as a civil society representative in CGDEX. While he was not happy the dam was being built, he embraced the opportunities to engage directly with the government in order to secure resources, and he remained allegiant to the Workers' Party. Those who remained opposed to negotiations and refused to accept resources also recognized the state's tactics. In fact, it was for exactly these reasons that leaders and active participants of Xingu Vivo complained about CGDEX and were disappointed by groups who took resources from the government. Regardless of which position each activist took, they seemed to be generally aware of state and elite strategies to maintain hegemonic power. Yet, because they had diverging opinions about how best to respond, they were entrenching the fractures in civil society opposition that effectively sustained the structures of control.

The democratic developmental approach to constructing Belo Monte may have weakened opposition and broken social and political networks, but it also opened new avenues of engagement and claims making that some dam-affected residents used to their advantage. As Antonio Gramsci's theory of hegemony suggests, the ways the dominant, pro-dam coalition was able to build enough consent to construct the dam and remain in control required constant work and struggle and was by no means permanent.[27] Subordinate groups, in this case dam-affected communities and their advocates, strategically maneuvered to resist domination and strengthen their own voices. In other words, these communities did not resign themselves to their situations; instead, they used the very institutions, like CGDEX and the law, that had once weakened their positions to seek compensation and justice. They formed new networks and developed new strategies. Both participants and opponents of participatory initiatives like CGDEX used those spaces as sites of resistance, as opportunities to exert agency, and as avenues through which to create meaningful democratic practices. The marginalized also leveraged the law—which the powerful had previously used to build consent to serve their interests—to creatively gain power and make material gains. How dam-impacted communities responded to state strategies and the challenges and opportunities they faced in confronting the democratic developmental state is the subject of the remainder of this book.

PART II AN ETHNOGRAPHY OF DAM BUILDING

4 · THE LIVING PROCESS

On March 22, 2013, I made my way to the monthly meeting of Comitê Gestor do Plano de Desenvolvimento Regional Sustentável do Xingu (CGDEX—Steering Committee of the Regional Sustainable Development Plan of the Xingu), the steering committee that brought together municipal, state, and federal government officials with regional civil society members to implement the sustainable development plan of the Xingu. By that point, I had attended nearly a dozen CGDEX meetings, so I knew what to prepare for. I expected to arrive slightly before the meeting started, find a seat in the auditorium that held nearly two hundred people, chat with a few people about topics that were on their minds, and settle in for a long meeting that would undoubtedly be filled with debates over both important issues and mundane topics. I had just returned to Altamira from a trip out of town so was surprised to see the familiar large banner of the anti-dam activist group Xingu Vivo that read "Justice Now." As I stepped into the foyer of the small two-story building, approximately fifteen people were gathered around Johannes, the federal official who facilitated the meetings and who would decide if and how the protestors could participate. I recognized many of the activists from Xingu Vivo and Movimento dos Atingidos por Barragens (MAB—Movement of People Affected by Dams). Given their disdain for CGDEX in the past, I was surprised to see them at the meeting. Antonia Melo da Silva, the founder of Xingu Vivo, and others were demanding loudly that they be let into the meeting. The group, which represented the neighborhoods that would soon be flooded and the families who would be displaced, wanted federal and state government officials to hear their concerns. After about fifteen minutes of pleading, Johannes let the protestors into the auditorium under the condition that only five of them speak once the meeting was underway.

We all made our way into the large room. Toward the front of the auditorium were thirty chairs facing the front of the room, with yellow police tape wrapped around them. They remained empty until the start of the meeting, at which point Johannes explained they were reserved for voting members. Johannes then turned the floor over to the protestors. Antonia began: "We are here to speak for the impacted neighborhoods. We have a right to be heard. This is, after all, supposed

FIGURE 4.1. Protestors at a CGDEX meeting that they stormed in March 2013. Sign on the left reads, "CCBM [Belo Monte Construction Consortium]—NESA [Norte Energia]—Government, When are you going to tell us the truth?" Sign next to it reads, "We want to *discuss* the resettlement areas." (Photo by author.)

to be a democracy, and CGDEX a democratic space. But people's rights are being violated!" She spoke forcefully for almost ten minutes, citing many problems with the displacement and resettlement plans and pleading for the government and Norte Energia to speak with the people. She concluded by directing her comments to other local residents and organizations: "Let's continue fighting! We have the right to negotiate!" She received a strong round of applause from many of the local residents and civil society representatives.

CGDEX had suddenly become a site of overt protest, but it had been a target of criticism and skepticism since the beginning. Activists who refused to participate in the space argued it was not truly democratic, that it would fail to support the communities most directly impacted by the dam's construction, and that it created divisions in the dam-opposition movement. These objectors were not alone in their complaints. Many local residents and representatives of nongovernmental organizations who regularly participated in CGDEX also recognized the limitations and challenges that CGDEX faced in accomplishing lasting sustainable development. These skeptical participants thought the funding was insufficient, there was no clear and uniting mandate that could actually bring about positive change and economic growth, and the people who most needed support would not be served by CGDEX, among other worries. Some participants even saw CGDEX as a state tactic to reduce resistance to dam construction. The skeptics who continued to engage in CGDEX were not shy about their complaints either, voicing them in discussions with me but also in the meetings. Why would they actively participate in a space they felt would provide few material benefits for the region and fail to accomplish its goals? What impacts would come from their participation?

Participants of CGDEX, particularly the representatives of local nongovernmental organizations and the dozens of citizens from the region, may not have stormed the meeting in an overt sign of protest, but they did use the space to wrestle for control and voice at every opportunity. For these participants, CGDEX was a space of subtle resistance where those with little other influence and power were able to subvert traditional power structures and exert their control and agency, making claims for their own futures and those of others in the region. Their ability to take some control was particularly important in a moment when they were experiencing the negative effects of dam construction and imagining a challenging future.

RECOGNIZING THE LIMITATIONS

CGDEX participants were well aware of the challenges of the participatory opportunity. Not only did activists recognize that the government might be using the forum as a political mechanism to gain more support for—and reduce resistance to—dam construction, but they also saw the financial limitations of CGDEX. The five hundred million Brazilian reals that Norte Energia had to provide the committee was a significant amount of money, but CGDEX was tasked with investing it in initiatives to bring about regional sustainable development in eleven municipalities, an area covering over sixty-four million acres of land with a population of over four hundred thousand people. Furthermore, the money was to be distributed over twenty years, leaving relatively small amounts for each year. Given the large geographic area, the significant population, and the complex challenges facing the region, many people felt the pool of money was woefully insufficient to create any lasting change.

This underfunding was a consistent topic of conversation in the plenary meetings. At nearly every gathering, somebody used this point to make an argument about the need to limit the projects they funded or how to divide the yearly allotment of money. After CGDEX had been meeting for a year, some participants began arguing that more of the overall funds should be allocated during the first six years, rather than distributing it evenly over twenty years. Dam construction was moving along, and the people of the region were feeling the impacts, so CGDEX should approve more projects earlier in order to support people impacted by construction. Some members also pointed out that the actual value of the allotted amounts would decrease due to inflation. Opponents argued this was a political move for more powerful people and organizations to acquire the funds. Elected officials might also take credit for investments. Soon after the debate began, the committee voted to shift the resource allocation so that more than half of the five hundred million would be used during the first six years. These decisions, however, did not increase the overall available resources or speed up the slow democratic decision-making process.

In addition to the financial challenges of CGDEX, participants had widespread concerns over the limited participation and representation in the committee.[1] It had a diverse set of members, but it was still limited to those chosen at the beginning of the process. This was a result of the structural design of the forum, the lack of interest from some sectors, and the lack of information about CGDEX among the population. The presidential decree that established CGDEX laid out the general makeup of the membership of the committee.[2] Thirty individuals representing organizations and government agencies would be official voting members. Each would be paired with a person from the same sector but a different organization or group who would serve as a substitute member if the primary member was not present at a meeting. Government officials would make up half of the members, evenly distributed between federal, state, and municipal levels. The five federal government officials were directors and managers of federal ministries and the Casa Civil, the executive office of the president of Brazil that coordinates, manages, and monitors government actions. The state government representatives were from state agencies, such as the secretaries of energy, health, and development. The mayors of the affected municipalities were the municipal representatives. The other half of CGDEX members would be representatives of regional civil society and divided into five sectors: four members from the business sector, four from urban and rural workers' and fishers' unions, four social movements and environmental organizations, two from Indigenous communities, and one from a teaching and research institution.

The presidential decree also indicated that representatives needed to be chosen within sixty days. With only fifteen civil society spots and another fifteen as substitute representatives, membership of regional organizations would be limited, and the process of choosing these representatives was unclear. Jonas, the entrepreneur who had helped form and then lead FORT, became instrumental in drawing up the list of civil society groups. He and most of the other members of FORT had been supportive of the dam, citing the potential positive economic and social benefits that it could provide if it were carried out in the best possible way. Jonas also recognized the negative impacts that could accompany construction and was deeply concerned that the potential problems could negate any of the benefits, so he actively participated in civic and political life in Altamira to try to help officials steer clear of these pitfalls. Jonas had been involved with the planning processes prior to member selection, and thus emerged as a representative of civil society in the eyes of government officials. The federal organizing body viewed him as somebody who could choose the members or at least devise a process through which those decisions could be made. This perspective of his leadership was not shared by everyone in the community, of course, so CGDEX participation was contentious from the start.

The groups who ultimately became official members did not fully understand why they were chosen, highlighting the lack of information about what CGDEX would be and who would have access to it. Some thought other groups were not chosen because of their lack of trust that the committee could be productive. Others credited their own participation to their long-standing relationship with Jonas or deep connection to the region. Eventually, some local leaders who were not chosen decided the forum was too heavily restricted and the selection process was unclear. Others were unaware that a new forum for decision-making was even created.

Participants also questioned who CGDEX would benefit. Sergio, the leader of a local group that supports fishers who catch and sell ornamental fish, or fish that are kept for aesthetic purposes, had a great deal to say about this point. I had met him at one of the first CGDEX meetings and saw him at a number of events throughout my time doing research. In April 2013, we scheduled a meeting at his office, located near the river's edge. He had told me to look for the building with flaking paint, as it lacked a sign of his union. When I found it, I walked through a front room that had a dozen or so Styrofoam coolers and a new freezer plugged into the wall. I poked my head into a small office in the back of the room and saw Sergio sitting at a desk. The room was just big enough for a single file cabinet, a simple desk, and chairs on each side of the table. A stack of white plastic tables and an old cooler on its side held a small fan in the corner. Another fan sat on the floor but was not running. There was no computer in the room and just a few stacks of paper on the desk. The slab of glass over the desk allowed me to see a couple dozen business cards advertising a wide variety of services and a smaller flyer from Norte Energia that had the phone number and an address of where residents could go to ask questions about the dam project.

As I sat down, Sergio was organizing paperwork and started complaining, "So many papers!" During our conversation over the subsequent hour and a half, he pulled out dozens of documents, reading sections to me that, more or less, aligned with what we were talking about. Since he knew I was interested in CGDEX, he almost immediately pulled out his copy of the *Plano de Desenvolvimento Regional Sustentável do Xingu* (PDRSX—Regional Sustainable Development Plan of the Xingu), which was designed to guide the work of the participatory steering committee. He read the title slowly, focusing on the word "desenvolvimento." He then exclaimed, "Development? Who benefits?" After a slight pause, he answered his own question: "The government. The only way society is benefitting is through school remodeling, some new police equipment, and other things like that. Society is suffering here. Health hasn't changed. Infrastructure is a big problem! Who's benefitting? The Indigenous, the federal government, the state government, and to some extent, the municipality." He went on to deride the current prices of fish and the government's lack of support for the fishers. When he mentioned "the Indigenous," he was alluding to compensation that Indigenous communities were guaranteed through mitigation programs. As we talked, he mentioned a few government projects that provided new training for fishers but said they were useless programs that shared information most fishers already knew. "We want compensation!" he said. He went on to talk about the lack of understanding of where the CGDEX money was going. He was particularly annoyed that CGDEX was funding projects that he felt the government should be carrying out and that should have been done years ago.

Later the same day, I met with a professor of agronomy, Carlindo, at the university in Altamira. I had met him a year earlier at CGDEX, where he was a regular attendee, but not a voting member of the general steering committee. He actively participated in the technical committee dedicated to supporting rural agriculture and was never afraid to share his opinions. At most of the monthly meetings, Carlindo and I had chatted with one another, so when I sat down with him in his small office crowded with books and papers, he seemed ready and comfortable to share his opinions of CGDEX and a host of other topics with me. He began talking extremely fast, using colloquial expressions, and deciding to say whatever seemed to cross his mind, rather than responding to what I was asking him. He had moved to the region with his family in the mid-1970s, when he was nine years old, and received his master's and doctoral degrees in agronomy from a university in the state capital of Belém. During his lifetime in Northern Amazonia, and particularly after working as a professor in Altamira for over ten years, he had experienced the lack of resources in the region and for his field of study. Despite the necessity of studying soils and crop production along the Transamazon, little money is available for the universities in northern Brazil. After sharing his dissatisfaction with the government support for agronomy studies in the region, he went on to complain about the lack of infrastructure. "It's our cancer," he said. "Roads, highways, waterways. The lack of this infrastructure is a big problem." The lack of basic societal services, in his opinion, resulted in what he saw as the foundational challenges of CGDEX. In a similar complaint as Sergio, he remarked, "Everybody wants a piece of the five hundred million [reals]. Civil society organizations and the government say they need it." He explained that the government cites the need for the money, even when they have other sources of revenue that are dedicated for projects that end up being carried out using CGDEX resources.

Carlindo continued by bringing up another point of contention I often heard: "The biggest sectors in the region are cattle farming, agriculture, and agronomy, but CGDEX is not talking about those things." He suggested this was a problem of voting. "The biggest sectors do not have more representation in CGDEX, so the smaller items get just as much, if not more, attention than the larger issues." Many people made a similar argument that CGDEX should have decided to use the resources for megaprojects, rather than dozens of individual projects. Cocoa, for example, could be a thriving industry along the Transamazon, but significant investments would need to be made first. CGDEX presented this opportunity, but nobody was organized well or fast enough to make that happen.

In 2017, after CGDEX had been operating for over six years and many of the structures and operations of the steering committee were well established, some participants came to share the same concerns as Carlindo and others from previous years. Gabi was one such person. She directed the local office of a nongovernmental environmental research organization based in Belém that carries out research throughout the Amazon. The organization strives to bring about sustainable development in the region by influencing public policy and implementing local initiatives. Gabi had been an active member of CGDEX since it began in 2011, voicing her concerns and ideas in meetings, while not embroiling herself in overly contentious issues. In 2017, she assumed a leadership role representing the civil society members. Through that position, she became more outspoken, working to support the array of expectations and demands of the many groups involved. When reflecting on CGDEX, Gabi saw positive outcomes of the participatory process, but, similar to Carlindo, she lamented what she saw as a lost opportunity to transform the region. She argued that instead of supporting relatively small projects, the money could have been used to develop the cocoa industry or strengthen the university. This type of investment could have created a development that balanced economic growth with environmental protection and social equity.

The dissatisfaction that Sergio, Carlindo, and Gabi felt was widespread. Every member of CGDEX with whom I spoke complained about the lack of a cohesive plan, the underrepresentation of key actors, or the inability of the committee to address the pressing issues raised by the dam construction. Despite their misgivings, nearly everyone continued to participate. In fact, people had invested considerable time and financial resources to participate for over six years. Some had to travel great distances and spend two to five days in Altamira every month. They spent energy creating project proposals that may or may not be funded. Considering the widespread recognition of the limits and problems of CGDEX, the continued level of engagement was remarkable.

PROCESSO VIVO

Participants stayed involved in CGDEX for a variety of reasons. Some organizations were attracted by the financial resources. Others had already been trying to shape the future of the region, and CGDEX opened a promising pathway through which to bring some kind of development to the area. More importantly, though, CGDEX offered an opportunity to be involved in what some participants called a *processo vivo*, or living process. In other words, people would shape CGDEX while participating in it. Other than the composition, little else about the functioning of CGDEX was determined prior to the first few meetings. Participants knew that the primary function of CGDEX was to allocate the annual portions of the five hundred million reals in ways that aligned with the sustainable development plan. *How* this body would do so was part of its mission during the initial stages of its existence.

The first two meetings of CGDEX took place in June 2011, just as dam construction began. They had opened the debate over the internal rules that would guide the work of the committee and determined that they would meet about ten times during the first year. Nearly all aspects of the committee remained undecided. The third CGDEX meeting took place in the first week of August 2011 and it began the work of deciding how the committee would carry out its broad mandate of instituting sustainable development in the region. It was the first meeting I attended, and I spoke with about a dozen people from civil society organizations in the month leading up to it. They all suggested that it would be important yet were leery of its purpose and its ability to create real change in the region. They also all warned me that I might not be able to enter, as they thought it would be a closed meeting in which only members could participate.

I needlessly feared that I would not be able to stay once the meeting started, so I arrived at eight a.m., a full hour before the meeting was scheduled to begin. The meeting would take place at the auditorium in the headquarters of the city's Commercial, Industrial, and Agropastoral Association (ACIAPA). I had just walked through the dusty and trash-laden streets of Altamira for a half hour and so was struck by the nicely manicured grounds as I approached the building. Less than a dozen people were in the auditorium, so I wandered through the small two-story building. In addition to approximately ten offices and a small conference room, the building had one of the few meeting spaces in the city that could accommodate close to two hundred people. As I re-entered the large room, a small group of men wearing Norte Energia lanyards were testing microphones and setting up a screen and projector. Another young man wearing a New York Giants hat set up a table with a sign-in sheet and a stack of badges. Each badge had a map of the region with the name of an official member of the group, whether they were from the government or civil society, and the specific organization the person represented. Later in the meeting, members would raise these badges when asked to vote. At the front of the room, a long table covered with a green cloth sat on a slightly elevated section of the floor. Hung on the wall behind the table was a banner that read "Juntos Somos Mais Forte" (Together We Are Stronger) with the logos of four groups: ACIAPA, the commercial union of Altamira and four surrounding municipalities, the association of retailers, and FORT Xingu.

About 150 chairs in rows of 14, with an aisle down the middle, faced the banner, and only a few people were seated when I arrived.

I sat down in the fifth row of chairs. As I waited for the meeting to begin, I was struck by the variety of people who were entering the room. A man walked in shortly after me wearing jeans and a T-shirt with the name of a local workers' union. A group of three women dressed in relatively formal attire sat together toward the front of the room. Three men in military uniforms sat in front of me, turned around, and introduced themselves to me. One of the men explained that he was the head of the battalion based in Altamira. After I had sat for about thirty minutes, people I recognized, who had told me about the meeting, began arriving. Leaders of two environmental nongovernmental organizations with offices in Altamira, as well as Simone and Claudio from FVPP, found seats on the other side of the room. Five men arrived wearing suits and ties, an unusual choice of clothing in Altamira, given the high heat, humidity, and agricultural culture of the region. I later learned they were from the state and federal government. The mayor of Altamira arrived shortly after them, just before the meeting started.

While I recognized some people there, I knew only a few of them. I had been in the city for a month, spending every day talking with as many politically engaged people as possible. Altamira was a small city, and I thought that I had, in at least some way, come in contact with the majority of the main activists and other active citizens. The people I had interviewed and with whom I spent time seemed to have represented diverse perspectives on the dam and were trying to support a variety of populations, such as farmers, fishers, and urban residents. The relatively few familiar faces may have reflected that my research had been limited, but it also showed that CGDEX was integrating new people into decision-making processes and bringing people together from throughout the region.

By the time the meeting started, approximately seventy people were in the room. Johannes, the representative from Casa Civil, facilitated the discussion. As he welcomed everyone to the meeting, his amplified voice echoed off the walls, making it difficult to hear. The mayor of Altamira then spoke for a few minutes. She talked about the importance of the moment for the development of the region and for creating a better life for people both here and throughout Brazil. A representative of civil society, as well as the state government, also gave a few welcoming remarks about the opportunities of CGDEX. Johannes then gave a brief presentation that outlined how and why CGDEX was formed.

After more than an hour of formalities and welcoming presentations, the serious conversations began. Since this meeting was early in the process, it laid the groundwork for how the space would function and for the topics that would become the most serious concerns and debates. It also showed how much of CGDEX was up to the participants, rather than predetermined. The first major debate revolved around how to allocate the annual allotment of money. One of the outspoken mayors made the self-serving argument that the fifteen million reals allotted for the first year should be divided equally between the municipalities. Representatives of civil society immediately countered his points, arguing that the money should be given to worthy projects, rather than distributed by a particular sector or place. Other mayors sided with the civil society representatives, pointing out that some municipalities, such as Altamira, were much larger and more directly impacted by Belo Monte. Ultimately, the proposal to divide the money by municipality was discarded.

The other major issue for the meeting was to create technical groups and decide who would serve on them. The presidential decree had stated that these small groups could be created to "implement the sustainable development plan of the Xingu and promote debates," but the composition, mandate, and role of these committees was left to be determined. The group began discussing all of these issues as they decided which committees to form and which were not in the purview of CGDEX. At times, these debates became heated, particularly when Eliana stood to argue that a technical group on health be formed. Eliana had moved to Altamira in 1960, learned to read from the Catholic Church, and became an activist as she fought against the dam and for better health care and other infrastructure. When dam construction began, she and her family nearly moved out of the region but decided to stay and fight to minimize the impacts. She was only slightly over five feet tall, but she had a commanding voice and powerful words that grabbed people's attention. People listened when she stood at the front of the auditorium, even though she was not a voting member of CGDEX. At this meeting, she cited the poorly funded regional hospital, lack of health posts in remote areas, and the rampant health problems as she pleaded with the group to focus some of their energy on health concerns. Others argued that health concerns would be discussed in other groups and that healthcare was not mentioned earlier in the process as requiring a committee. She and her supporters responded that CGDEX was an open process and that health was one of the most pressing issues of the moment. By the end of the debate, it was clear that the large group was not going to approve the new technical committee on health, but Eliana had opened up a conversation that would continue in CGDEX for years.

As the conversation over technical groups continued, I was astounded by just how much was *not* decided prior to the meeting and how many small details the group discussed. They agreed upon four technical groups quite quickly, as they were outlined in the development plan that led to the creation of CGDEX, but they discussed the details of what each would do at length. A group of Indigenous leaders and nongovernment organizations that supported rural extractivists, more broadly, pushed for a technical group to focus on issues related to those communities. Extractivists in this context referred to traditional communities that harvested and gathered natural resources from the forest, such as nuts and basic crops. Unlike the question on health, this suggestion garnered quick support. Par-

TABLE 4.1 Technical Groups of the Steering Committee of the Regional Sustainable Development Plan of the Xingu (CGDEX)

	Technical group	Main functions	When initiated
1	Spatial Planning, Land Regularization, and Environmental Management	Addresses complicated situation around land rights in the region	In original plan
2	Infrastructure for Development	Attends to infrastructure needs not considered in official mitigation programs	In original plan
3	Fostering Sustainable Productive Activities	Supports rural agricultural activities	In original plan
4	Social Inclusion and Citizenship	Supports education, women's rights, and other social justice concerns	In original plan
5	Monitoring and Accompanying	Examines and observes the dam's mitigation programs and projects	Added during initial CGDEX meetings
6	Indigenous People and Traditional Communities	Supports traditional communities, including Indigenous, fishers, riverine people, and others who live predominately subsistence lifestyles	Added during initial CGDEX meetings
7	Health	Addresses health needs in the region	Added later
8	Education	Addresses education needs in the region	Added later
	General Coordinating Body	Provides directives and organization of the steering committee	Formalized later

ticipants then discussed the name of the group for what seemed an exceptional amount of time. Instead of "Indigenous People and Extractivists," it was changed to "Indigenous People and Traditional Communities."

The debates over the makeup and function of the technical committees continued in subsequent meetings. Would projects be submitted to the general body or through the technical committees? How should money be distributed among the different Technical Committees in the fairest manner? Who can submit proposals to the CG? Who should be allowed to vote on the project submissions? The fight for a technical group dedicated to health care issues also continued until early the following year, when a group led by Eliana succeeded in convincing the committee it was necessary. The committee then dedicated a special allotment of money because of immediate health concerns. A technical group committed to education was later added, as well. Table 4.1 provides a complete list of technical groups and shows when they were created.

Each technical group was comprised of at least twelve voting participants, some of whom were CGDEX members but many were not. Federal representatives of CGDEX coordinated each technical group, serving as liaisons between the committees and a general coordinating body. In the end, the general committee decided that these technical groups would carry out much of the work of CGDEX. Members of the groups communicated between meetings, met the day before the general plenary, and then reported to the larger body to share the decisions they made, challenges they faced in carrying out their work, and ideas that they felt needed to be discussed by the full committee. More debates and discussions would then ensue in CGDEX.

Deliberative, participatory democracy was at work. CGDEX was deliberative because participants were engaged in meaningful, consequential discussions and debates in the public sphere. The space was participatory in that it allowed citizens to be involved in direct engagement and decision-making with the state in ways that ensured government officials would respond to their concerns, ideas, and perspectives.[3] The extensive discussions and decisions about the technical groups highlight just how fundamental participation and debate were to the functioning of CGDEX. Indeed, CGDEX repeatedly manifested itself as a living process being created by the people involved in it. The main body discussed nearly every aspect of their process. On what criteria should proposals be judged? What information would they request in the call for proposals? How should each question on the form be worded? Debates around these questions, among many others, would drag on for hours, often pushing the meetings late into the evening. There were even hotly contested discussions over the schedule of meetings. After decisions were made, members often returned to the same points at the subsequent meeting or after the decisions were implemented. Very few decisions, if any, were permanent.

The participants of CGDEX highly valued this open, participatory space that was not predetermined by the state and other powerful actors. From the start, participants from all sectors, including the state and civil society, wanted to protect the democratic values of the forum and celebrated "open participation" during the meetings and in one-on-one interviews with me. After the morning session of the first meeting I attended, a representative of a group that supports traditional communities seemed excited about what had been happening and the opportunities to have real conversations about issues that mattered. "These kinds of spaces are new in Brazil and they are trying to use CGDEX as a real democratic space," she explained and said she was looking forward to the discussions after lunch. "This morning was about politics. People were polite and respectful but just wait for the afternoon. That's when we will have real fights!" She and others embraced the difficult discussions, as they felt they were engaging in true democracy. When debates seemed to be getting out of hand or people claimed they were not being heard, Johannes, the meeting's facilitator from Casa Civil, would remind everyone that CGDEX was committed to the democratic process. Voting was a regular occurrence and nearly always sparked a disagreement over who could vote and if the votes were tallied correctly.

People cared about the democratic process, and they explicitly discussed it as such. A meeting in August 2012 highlighted how people invoked the "democratic process" in heated debates. The committee had received the year's round of project proposals and were discussing whether proposals with missing paperwork would be considered. The debate largely fell into two sides. Some people began by arguing that the rules were clear and that only complete proposals should be accepted. In response, Eliana, the nonvoting member of CGDEX who had been influential in creating the technical group on health, got out of her chair, grabbed a microphone, and walked to the front of the room. "You can't limit development to the thirty members of the steering committee," she began in her commanding voice, and then launched into a long argument as to why the process needed to be as open as possible. She concluded, "My argument is that all project proposals submitted should be accepted and evaluated." In response, a local leader of a construction business stood up and sounded annoyed as he remarked, "There were three meetings where we discussed this and decided on how things should be done. This was a collaborative process and you can't change the rules now in the middle of the process." Eliana and others responded that the rules were not, in fact, clear and that the process should continue to evolve. This debate went on for over an hour. At the core of the dispute was whether the decisions that were previously made through a democratic process should override the ongoing democratic process. At the end of the discussion, the committee voted, in an overwhelming majority, to allow all projects to be considered. Eliana had a huge smile on her face as she quietly celebrated with the leader of a local union who was sitting next to her. This was a victory for those who believed that CGDEX should remain as inclusive as possible and continue to be the "living process" that everyone applauded.

Given the ways these discussions unfolded, participants usually felt that the committee was succeeding in creating a democratic process. The main agenda item of the penultimate meeting of 2012, which took place just three months after the debate over project proposals, was to reflect on the year's proceedings, comment on the positive aspects of the process, and generate new ideas to better CGDEX for the following year. Each technical group discussed these items on the first day of the two-day meeting, generating lists of both the positive aspects and the things that needed improvement. On the second day, the groups reported to the general plenary. As a representative from each of the seven groups stood to provide a summary of their discussions, the audience paid particularly close attention. The first speaker, a representative of the federal government working in the technical group on land regularization, began his report by saying, "We had an intense debate about the positive and negative things from the year." He went on to talk about the wide array of project proposals and the fact that new spaces were created through which people could discuss them. "The biggest positive point is that we have learned a lot from the process [in the past year] ... Everyone thinks some projects are good and others bad, and that is a result of the democratic process." The other reports began with similar references to their productive discussions in reflecting on the year of work. Many of the others' positive reflections of the year were also about the democratic way in which the process was carried out. Eliana, speaking for the technical group on health, reported, "The important positive thing was using a participatory democratic process. Since creating the call for submissions, this process [of selecting projects] was done in a democratic and participatory form."

Nearly everyone in CGDEX was employing the terms "democracy," "dialogue," "open participation," and the "living process" as they engaged in discussions. Most used them in an attempt to better the space, by making implicit and explicit arguments about what CGDEX stood for and what it should be in the future. In addition to the federal and state government officials, many non-state and local participants also believed in the process. Even if they did not think that the region would see a great deal of material gains from CGDEX, they felt empowered to engage in a democratic process that allowed state and non-state actors to deliberate on what felt like more equal terrain. This deeper form of democracy, for them, had long been missing from the region and thus many people welcomed it. The more they engaged with the living process, the more they believed in it.

This belief in the process made CGDEX empowering. Scholars suggest participatory institutions are empowered when the structure of such spaces "attempt to tie action to discussion."[4] But for many of the participants, the discussion—insofar as the discussion was deliberative in nature—was itself an action. For some observers, this might be seen as a form of weak empowerment, because the process mattered more than the outcomes. For participants, though, they felt empowered because CGDEX offered them an opportunity to be full participants in a decision-making body. Even if their demands were not fully met, they believed that the political actors to which they had access and the participatory institution in which they were engaging were responding to their claims.[5]

The participants' celebration of the democratic process served to justify their continued participation while they were also dissociating themselves from support of the dam. Skeptical participants kept their distance from the negative effects of dam construction, corruption, and even the problematic aspects of CGDEX by applauding the process they saw as participatory and empowering. While enthusiastic participants cited the idea of the living process as a means to justify the existence of CGDEX and even the dam itself, skeptical participants' embrace of the same rhetoric allowed them to justify their decisions to participate while remaining critical of the space and the dam.[6]

DISPUTING DEMOCRACY

While both enthusiastic proponents and skeptical participants lauded the democratic practices of CGDEX—even as they employed it with different goals—activists like Antonia Melo and Irá, who consciously refused to participate, overtly criticized those same practices. The leaders of Xingu Vivo and MAB were frustrated by the ways the federal government used CGDEX to claim Belo Monte was being built in a democratic and participatory manner. For these activists, CGDEX was simply a way for the government to quell resistance to dam construction and co-opt alternative ideas for development. The argument that CGDEX was a democratic space further annoyed them, as they believed that the kind of dialogue that took place in CGDEX was not real dialogue. They felt that the opportunity to participate in this type of decision-making process was a weak form of empowerment at best, and destructive at worst. The anecdote that opened this chapter highlights how these activists viewed CGDEX but also what they were demanding in its place and the ways they were attempting to reframe the discussion. When they forced their way into the meeting in March 2013, they showed that their disdain for CGDEX did not stop them from using the participatory space as a target of their resistance and as an opportunity to tell a different story. They were attempting to rescue more radical and emancipatory visions of democracy from being lost.[7]

By early 2013, less than two years after dam construction and CGDEX meetings began, the resettlement process of soon-to-be displaced urban residents was becoming a central concern for activists and many of those residents. The forced displacement of nearly eight thousand families living in Altamira would begin about a year later, but most people were still confused about what would happen. Norte Energia had announced plans to build concrete houses on the edges of town in collective urban resettlements and were holding large neighborhood meetings to discuss the resettlement plans with community members. Activists, some residents, and outside observers derided the announced plans, arguing that concrete houses would be unbearable in the heat and humidity of the Amazon and inadequate for many of the families who would be displaced. Furthermore, many people were unhappy that the resettlement communities were far away from the center of town and without adequate transportation options. Xingu Vivo, MAB, and other activists began developing strategies for how to make demands around housing issues. These groups attended the neighborhood-based information session to raise their concerns but felt frustrated by the lack of response.

They then made CGDEX a target. Six weeks after the first time the social movements had forced themselves into the CGDEX meeting to voice their concerns about the resettlement process, I stopped by the offices of Xingu Vivo. Their office was teeming with energy and action. As I walked up the stairwell, I could hear people debating what a house should look like and laughing with each other as they argued over colors and designs. I entered the open space where various organizations had coolers, posters, tables, and chairs along the walls. In the middle of the room were Irá, who was one of Xingu Vivo's principal organizers, and two younger women building a cardboard house. I had known Irá for nearly two years by that point and had met the other women earlier in the week, so they immediately welcomed me with smiles and explained that they were constructing the ideal home that should replace the awful concrete homes that Norte Energia was planning to build.

In the adjoining room, Antonia, a photographer from São Paulo, a Greenpeace representative who worked in another Amazonian city, a German researcher, and a dozen other local activists and journalists from at least three different organizations were talking with each other in small groups and working on computers. I had been to these offices dozens of times. It was usually a quiet space, with one or two people working and maybe a few others cooking and talking calmly. I knew that the level of activity and number of people in that space meant they were planning an event, responding to a crisis, or organizing logistics for a protest. I asked Antonia what all the activity was about. She explained that they were supporting an occupation of the dam site that Indigenous groups had just begun, but that was not all. They were also planning a big rally the following day, which was when CGDEX was scheduled to meet. I began to ask her if the event would be at the CGDEX meeting, and she put her finger over her mouth to indicate it would be a surprise for many of the CGDEX participants. After the small group had stormed the meeting less than two months earlier, they were afraid the government would prepare a blockade if they knew about the protest in advance. She handed me a flyer: "Protest Against the Inhumane Housing Project for the People of Altamira." The flyer invited residents of the affected neighborhoods of Altamira to come together to demand dignified and just housing.

While I read, everyone took seats around a large table to discuss their plan. Antonia began by explaining why they were going to CGDEX and what they would do: "One of our points of attack is [CGDEX]. People say 'Let's go close the street' but who is going? What would we do there? Let's go to CGDEX and demand that the presidents of Norte Energia and the high-level federal government officials come talk with the people. Let's demand that [the coordinator of CGDEX] do this, to arrange a public hearing." Antonia explained that they were targeting CGDEX because of the dozens of federal, state, and municipal government officials who would be forced to listen to their arguments. Despite their refusal to participate in CGDEX, the presence of federal representatives, in particular, made the meeting an appealing target. Antonia and the others felt the federal officials at the meeting actually had the power to fulfill their demands.

By the time I left, the model of the "ideal home" was nearly finished. Blue walls were painted with the black outlines of rectangles that represented bricks, a direct challenge to the plans to construct concrete homes. The lower edges of the model were covered by green paper that looked like grass, and a white door and open window made it look inviting. In the window was the figure of a man with his arms folded. A speech bubble next to the man spoke for all of the activists and residents concerned with the displacement plan: "We demand dignified housing!" The group had drawn shingles onto the brown roof of the structure. In each shingle on one side, they wrote the name of a neighborhood from which people would be displaced. On the other side, they inscribed the roof with two slogans: "Justice Now!" and "Comply with the Conditions!"

The next day, over one hundred protestors brought this house, dozens of signs, and their voices to the CGDEX meeting. They had gathered a few blocks from the CGDEX meeting place and marched to the entrance of the building. The protesters were met there by the facilitators of the meeting, who were reluctant to let the boisterous crowd inside and insisted they talk outside. Two women wearing MAB shirts hoisted the model home in the air, chanted, and demanded to enter, which the facilitators eventually allowed them to do. The large group filed into the meeting hall and placed the model home at the front of the room so everyone could see it. They set a MAB flag next to the home, and off to the side hung a larger banner that read, "President Dilma, we demand the cancellation of the concrete house plan!" and was signed "Residents of the Neighborhoods of Altamira." People in the audience held signs that made similar points: "Dilma government, how long will you keep lying to the people?" "Concrete homes are a crime!" "Dilma, we demand the cancellation of the inadequate urban resettlement program!" These signs directly addressed the federal government, whose representatives were part of CGDEX.

For the next hour, over a dozen people from the protest group spoke about their demands. They complained about the lack of information, the indignity of concrete homes, the lack of respect for the people who were being affected, and the ways they felt the government was violating the Brazilian constitution. They also spoke directly about democracy and how CGDEX was anything but an open participatory opportunity. "Dialogue is like this: one side talks, the other side listens, and then you come to a conclusion," Antonia began. "We need to have a *real* democratic dialogue. You need to change your style of dialogue." Leaders from MAB and residents facing imminent displacement from their homes made similar points. They explained that there was no real discussion in CGDEX or elsewhere to hear from and address the needs of people directly impacted by the dam.

Just before they left to allow the meeting to continue as previously scheduled, Antonia made closing remarks directed at each sector of the CGDEX members: "I appreciate that the people from the population and civil society are here. You have a responsibility to the population here. They are only receiving bad things from this project. You have a responsibility! And to those of you from the government—you have to reflect on what is happening. You need to evaluate. You need to think about the democracy that *we* want. We aren't being respected. We demand a public meeting in which everybody can participate. If you don't comply, we will continue fighting to your face."

The complaints about the type of democracy used in CGDEX was not lost on the government officials, who pushed back against the activists' criticisms, but also agreed to their demands. After Antonia finished speaking, Johannes began his response by saying, "We have norms and laws to have a democratic process." He then suggested a date for the public meeting three weeks in the future and said that they would continue to respect the process. As he concluded, he insisted that the government and CGDEX were listening and maintaining a democratic and open approach.

THE CLASSROOM OF DEMOCRACY

Enthusiastic proponents of CGDEX, skeptical participants, and even conscious objectors were arguing for democracy, dialogue, and open participation, but from quite different perspectives and for varying reasons. The activists who targeted CGDEX as a site of protest contended that the discussions in the meetings did not constitute real dialogue. Representation was limited, and, in their eyes, the populations impacted by the dam were not benefitting from the work of CGDEX. These objectors did not believe that the work of CGDEX would lead to material outcomes that would benefit the most marginalized and at-need communities. The protestors used the term "democracy" in attempts to strengthen their arguments and discredit the work of CGDEX.

The enthusiastic proponents made the counterclaim that CGDEX was employing robust democratic practices and that the open participation was a success in and of itself. Many of these advocates of CGDEX were state actors, particularly from the federal government, and other people representing organizations and agencies outside of the Transamazon region. They argued that CGDEX was an important space for the region's residents to learn about civic engagement and collective thinking, while downplaying the need to obtain material outcomes. Some government officials had experience in other participatory processes and thus had low expectations about CGDEX, focusing on the process as the end goal. Lucas, who was the director of the federal government's energy programs, had been the secretary of public works in the southern city of Porto Alegre for sixteen years, during which time the city was the first to institute participatory budgeting. I interviewed Lucas in his office in Brasília, where he explained the difficulties facing CGDEX and why the democratic nature of the space was vital:

Nothing in CGDEX surprises me. It is interesting but the challenges and changes don't surprise me. This is part of the group process ... In my experience in Porto Alegre, I learned that the process is the thing that is most important. People learn how to discuss with others. In Altamira, you never had this. So, this is the initial phase. It is a phase of showing this democracy. Things will change. It will not end. The value of learning how to participate is something that will continue. That is something that has been a big benefit for the region. And now it's there.

Felipe, a member of CGDEX and an architect who had worked in various capacities with the state government of Pará for thirty years, also felt the introduction of CGDEX had benefits for improving discussions around important issues facing the region.

People are not used to this kind of collective thinking. It's difficult to do. The municipalities and their mayors as well have a hard time thinking about the region as a whole. Instead, they think about their individual municipalities. Unlike the state government, people have never had to think more broadly. Now, for the first time, there is an attempt to talk to the people being affected. [CGDEX] is an experimental program. It will not remove the damages [of the dam], but at least people are starting to think about the environment, society, and the other things that are being affected ... People are not accustomed to the technical details and sometimes don't know what is being said but this is what needs to happen. It's all an experiment and these discussions are good for everyone.

Lucas, Felipe, and other enthusiastic proponents saw open participation, dialogue, and collective thinking as the end goal, rather than a means to more material benefits. By focusing on the importance of the democratic process, they often devalued outcomes that would better serve marginalized communities and those impacted by dam construction. Their enthusiastic celebration of the democratic process justified inaction and the slow process of CGDEX.

Skeptical participants, including everyday citizens, activists, and representatives of local and regional organizations, celebrated the democratic process in similar ways. They were more critical of the lack of material gains from CGDEX than the enthusiastic proponents, but these participants also lauded the ways people learned how to debate, think about the collective, and engage in difficult decision-making processes. The representative of the Sindicato de Trabalhadores e Trabalhadoras Rurais (STTR—Rural Workers' Union) in the region who participated in CGDEX was one such person. After he complained about the lack of any concrete and significant regional changes, I asked him if there were positives aspects of CGDEX. "Of course," he replied. "In particular, this is the first time we have participation of civil society [in decision-making]. It's a space to debate the effects of the dam. This is the most democratic space we have and it's in this big project. People are learning to debate.... This is a new process, a new way to think. With the Tucuruí dam [built in the 1970s–1980s], for example, the effects were never discussed with the people and they still haven't been."

The praise that skeptical participants gave CGDEX sounds like the praise of the more enthusiastic ones, but it differed in consequential ways. Not only did the skeptics more directly address the lack of material benefits, but they also used the language of democracy to assert power and agency in otherwise unequal environments. Gabi, the steering committee's representative of civil society members, argued that there were important benefits of the participatory opportunity: "CGDEX is a school teaching people how to transform ideas into projects. In particular, it is a space that is helping people articulate ideas to a wider audience." While this was similar to how state officials celebrated CGDEX, Gabi suggested that the benefits went beyond residents learning how to participate in collective thinking. For her, CGDEX aided people in developing the skills and knowledge necessarily to be empowered political citizens. "It forces people to share their ideas and engage in debates," she said. "It is a school of political construction."

Skeptical participants recognized that people from the area were not prepared to engage in debates and discussions about regional development. The government's decades-long inattention to their well-being meant that many people lacked the human capital—or skills, education, and knowledge—that many of the federal government officials possessed.[8] Less than 30 percent of people had an eighth-grade education or above.[9] In addition, while some people in the region had effectively organized to make demands on the state, the form of deliberative decision-making in which they were engaging with CGDEX was entirely new. These conditions did not prepare many people to be equal participants in CGDEX, particularly given the inherent power differentials and the experiences that federal and state officials acquired in previous experiments with participatory democracy.

Despite this lack of capital and insufficient preparation, many residents and representatives of local organizations used CGDEX as a site of resistance and an opportunity to exert agency. They did so in part by using the language and practices of democracy to assert power in CGDEX itself and improve the space. The debate over whether to accept project proposals provided a good example. As Eliana argued for an ongoing democratic and open process to accept more project proposals, even those that did not comply with regulations, for example, she found a strategic way to use the foundational principles of the space to benefit her argument. Even though she was not an official member of CGDEX, she employed the celebrated language of democracy in order to win arguments and create a more inclusive space. Moments like this highlight how skeptical participants reasserted agency in CGDEX. While many enthusiastic proponents in CGDEX had employed democracy to justify inaction, those who were usually in subordinate positions of power adopted similar language to demand more from the experience, make claims, improve the participatory space itself, and fight power imbalances. Not unlike the overt protestors of CGDEX, many skeptical participants used the space as a site of resistance, claiming all the power they could.[10]

In sum, the focus on the democratic process and the benefits of open participation meant different things based on the position and sector from which people were speaking. Rather than view CGDEX and other participatory opportunities as co-opted or as failures when they did not lead to material benefits, we can see that these initiatives might have provided other benefits that were consequential to democracy.

SHIFTING MINDS AND RECONFIGURING NETWORKS

As CGDEX became a site of overt and subtle resistance, people's views of collective action and decision-making began to change. CGDEX brought together a wide assortment of groups with diverse interests for the first time. As detailed in chapter 2, from the 1970s to the 2000s, regional social movements, the Catholic Church, and the Workers' Party had organized residents in the region to fight for basic infrastructure and rights, but this mobilization was limited to segments of the population who shared similar challenges. CGDEX was different, because it created discussions across different social classes. As the discussions of democracy in CGDEX show, these open conversations encouraged people to move beyond the corrupt, self-interested nature of local politics that they felt had been a hurdle in solving problems and improving living conditions in the region. According to many participants, CGDEX became one of the first spaces where people thought about the region as a whole, learned how to make collective decisions, and mobilized across different sectors. In so doing, CGDEX led to new, unique, and surprising alliances.

The story of Sonia, the longtime head of a union that supports construction workers and a member of CGDEX, captures participants' widely held view of CGDEX as an important forum for collective thought and the new partnerships that may form as a result. In the CGDEX meetings, Sonia was soft-spoken and always polite but asked pointed questions. She listened carefully to the dialogue during the meetings and was interested in hearing viewpoints that were different from her own. In May 2013, she and I met for an in-depth conversation in her simple, clean, and well-organized office. An air conditioner hung on the wall and her desk had a computer and a pile of papers neatly stacked in the corner. Sonia welcomed me into her office with a smile, offering me water and coffee. As I sat down, I mentioned that I was living in an apartment just down the street. She said that her cousin was my landlord, and he also owned an auto-parts store. She talked about how she had worked in that store for a short time when she arrived from the southern state of Paraná in 1999. She knew nothing about cars and disliked the work, so visited the union office, which was in another part of town at the time. The president of the union offered her a job cleaning the building and providing basic day-to-day assistance for the workers. After two years, she became the secretary of finances for the union, which at the time supported workers in the forestry sector. Soon after she took this management job, the union decided to move their offices to the center of the city. "Our first order of business was to build a new office," she said. "We had to really fight for this space. After we managed to buy the land, every cent we made went to construct the building and buy furniture and other things we needed to run the union from here. This was *really* important." Five years later, she was elected president of the union, and she proudly recalled that she received 80 percent of the vote.

At the time of the union election, 70 percent of the forestry companies whose workers were represented by the union were in the midst of closing, putting thousands out of work. The companies, workers, and union blamed the crisis on new Brazilian regulations to limit deforestation. Around the same time, it became clear Belo Monte would be built. With so many people out of work, she and most of the eight thousand workers the union represented saw hope in the dam. Belo Monte could offer thousands of new jobs and opportunities to train workers in new areas, which they could use after dam construction was completed. They also recognized that, along with the dam construction, there would be dozens, if not hundreds, of smaller projects that would need workers.

Given this context, it is not surprising that Sonia was instrumental in organizing a rally *in support* of the dam in 2006. She and the thousands of people who flooded the streets wanted to show that there was a great deal of backing for the project, not just opposition from vocal anti-dam activists. They also wanted to participate in the process. "We were demanding to be a part of the work," Sonia explained. These desires to provide labor for construction and engage in larger decision-making processes carried over to her participation in other ways. She made sure to be active in all the public meetings about the dam, even traveling to Belém, the state capital, to attend important meetings. "If we close our eyes, what happens?" she asked rhetorically when talking about paying close attention to all of the participatory opportunities.

This active engagement led to the union's participation in CGDEX. After the rally in support of the dam, she helped form FORT Xingu, the collective of over 130 business entities that Jonas would eventually lead. According to Sonia, when Jonas was choosing civil society representatives for CGDEX, he selected the union because the workers it represents know the history and life of the region. "They have lived the suffering," she said.

Like other participants in CGDEX and residents in the region, Sonia once had a distrustful and pessimistic view of fellow citizens and politicians as purely self-interested. Sonia and others thought CGDEX was playing a role in changing this: "[CGDEX] is a point of unification, a place of opportunity for people to come together, and people are starting to actually think collectively." Although she lamented the slow progress and bureaucracy of CGDEX in comparison to the rapid construction of the dam, she felt that "changing the way people think leads to getting things done." In her opinion, CGDEX was encouraging people to think about development on a regional level, rather than just about individual projects or municipalities. After two years of participating, people were beginning to think and act with the collective good in mind.

Not only was collective thinking advancing, but groups were also creating new, long-lasting partnerships. Sonia's union, for example, developed a project with FVPP, the foundation representing small farmers that had long opposed Belo Monte, and its leaders Claudio and Simone. Members of both the union and organizations cited the mutual challenges facing the region as a reason for coming together to develop projects that could support the distinct but related populations they aimed to serve. They explained that CGDEX made this partnership—and others like it—possible.

Both of the protests carried out in CGDEX meetings by nonparticipating social movements also led to surprising alliances between people like Sonia and staunch anti-dam activists like Antonia Melo. The messages brought to CGDEX through these activists resonated with participants like Sonia, even though she had once rallied in favor of construction. A few weeks after the first time Xingu Vivo barged into CGDEX and a month prior to the rally with the model home, I attended a meeting organized by Xingu Vivo in which they called a range of groups together to generate support for actions against the dam. Xingu Vivo, seen as quite radical by many city inhabitants, had been finding it difficult to organize large rallies or even smaller groups to support their causes. It was thus surprising to see nearly fifty people representing an assortment of governmental and non-governmental groups, and particularly members of groups who had historically rallied in support of Belo Monte, such as Sonia.

In my interview with Sonia, she explained why she attended: "I had supported dam construction for the promised work opportunities for the people, but I also see the immediate need to demand that the conditions to mitigate the negative impacts of the dam are met." She went on to say that she had preconceived ideas about Xingu Vivo and did not understand what the movement wanted. After the first protest at CGDEX, she changed her mind about the group. "Xingu Vivo fights are the same as ours and they need to be strengthened. People need to come together. It was at CGDEX that I learned more about the movement. But we can't wait for CGDEX or the mayor." For Sonia and others, CGDEX is not necessarily where all of the progress happens, but it is the place where people can find common ground, build new partnerships, and create strategies to accomplish their goals. The first Xingu Vivo protest at CGDEX had catalyzed new alliances, which helped form a larger coalition that the activists mobilized to carry out larger events, such as the second rally that had a much larger turnout and received some support within CGDEX.

Some of these alliances were short-lived, as they centered on specific issues. In other ways, though, the shifting mindsets and views toward other groups were long-lasting. Gabi, for example, argued that it was very important to bring diverse groups of people with many different interests together in this type of participatory space. Historically, she recounted, people here had many preconceived notions and prejudices about other groups. Commercial groups and unions, for example, thought that her environmental institute was opposed to development. Her environmental institute viewed those commercial groups as indifferent to environmental and social issues. According to her, six years of working together in a "democratic space" forced people to talk with each other and understand different perspectives. This had broken down barriers and reduced prejudice.

FROM CONFLICT TO DELIBERATION

Shortly after dam construction began, the era of rallies and protests that divided society into two camps of "for versus against" Belo Monte was largely over. Money, new opportunities, and new people poured into the region. Whether previously in favor or opposed to the dam, organizations and engaged residents had to navigate the new resources to make demands, organize, and work to reduce the negative impacts of construction. CGDEX became central in this reconfiguration of political contestation and moved at least part of the conflict from a two-sided conflict to a deliberative decision-making process.

Many of the people who had been anti-dam activists for decades understood that the government could, and perhaps was, using CGDEX to co-opt their interests and demands, weaken collective mobilization efforts, and restrict their ability to make claims in other ways. Traditional forms of resistance had indeed been upended, but CGDEX was not only destructive. It was also productive. New alliances formed, people learned new models of active citizenship, and the deliberative, participatory space became a site of resistance against traditional structures of power.[11] As participants engaged in CGDEX, whether through active participation or overt criticism, they used the space as a form of opposition and to establish power and exert agency.

5 · THE FIGHT FOR RECOGNITION

In early December 2017, six years after construction of the dam began, I returned to Altamira at an opportune time. It was the sixtieth birthday of Gloria, one of the activists I had met at Xingu Vivo's office six years previously. She was a vocal opponent of the dam who had lived along the river for decades but who had been forced to move due to construction. I remembered Gloria, in part for the way she spoke incredibly fast and also for her determination in fighting for justice. Because of her rapid speech and her accent, I needed to pay close attention in order to understand what she said, which was often something critical of the government or Norte Energia. She always seemed to be angry about Belo Monte, but also jovial around her fellow activists. Gloria refused to tolerate injustice of any kind and was not afraid to confront those in positions of power.

Coincidentally, Gloria's birthday fell exactly one year after the official creation of the Conselho dos Ribeirinhos, the Riverine People's Council. The group was made up of people who had lived along the river and had been forced from their land due to dam construction. They fought to be fairly compensated and to be relocated back to the river. Gloria was one of the first of the group to be granted land on the river, and she had already built a house and other basic infrastructure. Given the situation, she thought it would be appropriate to celebrate her birthday at her house and also celebrate the victories that the riverine people had achieved.

On the morning of the celebration, I went to the central river port—which consisted of a paved ramp to the water's edge and a couple dozen boats of various sizes—to catch a ride to the festivities on Gloria's land down the river. When I arrived at the shore, a dozen familiar faces greeted me, as they waited to board a long skinny metal boat. Antonia Melo da Silva, the leader of Xingu Vivo, was there, as were some journalists and staff of nongovernmental organizations whom I had met in the two weeks since arriving. I also reunited with some of the people I had come to know quite well during my previous research time in Altamira, such as the local leaders of the Instituto Socioambiental (ISA—Socio-environmental

Institute). A few of the fishers whom I had spent a great deal of time with along the river and in meetings in Altamira were also there, prepared to bring everyone to the party.

I boarded the boat of a fisher I knew from previous protests and felt fortunate to sit down next to Thais Santi, one of the federal prosecutors who worked in Altamira. Thais played a central role in the riverine people's struggles but was extremely busy and often traveled to other parts of Brazil. The boat ride would allow me to hear her perspective on the situation surrounding the fishers and riverine people. Thais and I met in 2012, shortly before the fishers' protest and occupation of one of the dam's construction site. When we met, she had just arrived to the region to assist with the expanding needs associated with Belo Monte. Her role as facilitator of the conciliatory hearing that brought the occupation to an end had concerned social movements, activists, and others fighting to support dam affected people. She quickly allayed these concerns, and in the subsequent five years, she became one of the key supporters of the fishing community and riverine people. Thais often had a warm smile and a friendly greeting but did not hesitate to speak frankly on important issues and push Norte Energia officials and government representatives to support the populations she represented, particularly when she felt they could help amend injustices. Thais devoted a great deal of time and energy to her cases. She had no connections to Altamira before she arrived, but over time, the communities she worked with embraced her and the hard work she was doing. She wanted to get to know them well and to understand their lives and challenges. This allowed her to more effectively fight for their rights and in support of their livelihoods.

During the half-hour boat ride to Gloria's house, Thais lauded the persistence and hard work of the riverine community in fighting for rights as a dam-affected population. She explained the long negotiations that the group had faced since the monthlong fishers' protest that I had witnessed over five years previously. I was astounded to hear the progress they had made, and she was equally excited to share their improbable story. She briefly explained how she had supported the group in creating irrefutable evidence of the suffering the dam project had caused and how she and the riverine people were negotiating for compensation. I was particularly surprised to hear that this was one of the only causes that had not, at any point, become a case with official legal proceedings. Instead, Thais and Andreia Barreto, the public defender, had served as mediators between the people who fished and lived along the Xingu and the targets of their demands, including Norte Energia and government officials.

The conversation was as enlightening as the scenery around us was depressing. We passed the newly barren island that was visible from the city's waterfront. The trees that remained standing had lost their leaves, and the bottom third (about six feet) of the trunk of each tree was visibly darker than the top two thirds. It looked like a long drought had stripped the life from the vegetation, but instead the reasons were quite the opposite. The new pattern of flooding from dam construction had drowned thousands of trees. We passed a beach that Norte Energia had created in the previous year, to replace the many natural beaches that were lost due to the construction of the dam. It was a beautiful Saturday morning, yet nobody was there enjoying the white sand and warm water. It appeared to be situated well outside of the city and surrounded by a barren landscape, so there was little appeal for local residents.

After the boat ride, we disembarked and walked about two hundred yards up a gradually rising dirt path to reach Gloria's property. As we stepped onto her land past two large mango trees, we passed a half dozen men who were tending to a large firepit and buckets filled with red meat. An open area about forty yards wide by sixty yards long sloped up to a two-story house, which overlooked the property. A large mango tree in the center of the area and rows of trees around the perimeter of the property provided an abundance of necessary shade for a row of simple benches that faced a long table filled with decorations of banana leaves, mangos, and other fruits. This table would double as the altar for an outdoor church service and eventually a serving location for lunch.

One of the Catholic priests, who had long been part of the fight for rights in the region, fully adorned in his robe, began the celebratory church service about an hour after I arrived. He and a small group of women who were active leaders of Xingu Vivo read scripture and led the group of about seventy people in songs. Both Gloria and the priest then spoke about the struggles they had faced and the accomplishments they had made, before providing opportunities for others to speak. From my perspective, it was an emotional moment. Most of the riverine people at this celebration were the same fishers who, five years previously, had protested at the construction site for over a month, until they could air their grievances at the conciliatory hearing. At that time, very few, if any, of the fishers had been involved in a political rally of any sort. In fact, very few had been civically engaged in any way. Dam construction changed that, as it upended their entire way of life. In the time since that protest began, this group of men and women had become active citizens who were finding ways to dictate their own futures at a time when others seemed to have much more influence over their lives.

Jose was one such fisher whom I had met in 2012. Born in the 1940s to a family that fished the Xingu, he had been on the river for nearly his entire life. His skin was dark and slightly wrinkled from decades of strenuous activity under the sun. When he walked, his back was a bit hunched over, but he appeared muscular and strong. He carried himself with a smile but was also quite serious. At meetings and other gatherings, he often dressed in simple polo shirts and shorts, but occasionally wore attire to symbolize his partly Indigenous background. A few days before the celebration, for example, I saw him at a meeting in which he carried a flat wooden club and wore a red beaded necklace and a wide, colorful bracelet. During the protest in 2012, Jose had been unsure of what to do. At the time, he spoke about his situation and the difficult times he was having since dam construction had begun, but he also told me that he had never done anything like that before and he did not think of himself as an activist. Like the others with whom he was protesting, he did not know what would come of it or what the next steps would be, but he explained that he was committed to continuing. Now, five years later, as they commemorated their successes, Jose seemed to be a professional activist.

Toward the end of the service, the priest invited comments from anybody who wanted to speak. Jose raised his hand and then stood from the wooden plank he had been sitting on. He began by wishing Gloria a happy birthday. He went on to say that they would continue to fight, but that today he was happy to celebrate with the people who had supported him and the other riverine people. As he thanked each person by name, including Antonia Melo, Thais, the various people at ISA, and his fellow fishers, he looked around to find them. As he did so, he became emotional, clearly caught up in the moment. He cut off what he was saying, looked to the sky, put his hands over his eyes, shook his head a few times, and began to cry. He was overwhelmed with gratitude and pride for what they had achieved.

The accomplishments that the fishers and their supporters achieved in those five years were indeed impressive. The protest and occupation in 2012 did not initially lead to financial compensation or other material benefits that the fishers were hoping for, and the group did not protest again, but it began a process that would eventually pay dividends. Through their protest, the men and women who fished the Xingu for a living had created opportunities to directly negotiate with the government and Norte Energia. They also used the protest, which was the first time most of them had participated in a demonstration of any kind, as a catalyst to become and stay actively involved in those negotiations. By the time they were celebrating on Gloria's property, the riverine people had managed to become recognized as an official council representing a group impacted by the dam, were promised a monthly stipend, and were hopeful that Norte Energia and the government would give more families land on the river. How, exactly, did this group manage to accomplish so much?

The story of this small group of men and women who fish and live along the Xingu River provides an example of how a marginalized group can overcome significant constraints in order to exert agency and to have a say in their own future. Previously left out of participatory processes, they nonetheless managed to access some of the resources coming into the region and create opportunities to engage in deliberations. They did so through protest, a great deal of persistence, and by mobilizing and activating parts of the state apparatus to work on their behalf.[1] Once the fishers had gained access to participatory opportunities and built their network of advocates, they began to gain ground in the conflict by expanding official understandings of who was impacted by the dam's construction and in what ways. In other words, these fishers had at first struggled to be seen as having legitimate knowledge. Their marginalized position allowed Norte Energia and government officials to disregard their opinions. But over time and with their supporters, the fishers were able to generate new understandings with evidence-based research. They also shifted the basic grounds of their claims making, changing their demands from being about what they did to where they lived. This story, in which dam-affected residents were able to make gains in the claims-making processes, shows both the systemic barriers to doing so and their agency.

THE FISHERS' PLIGHT AND RESPONSIBILITY

When construction of Belo Monte began, the fishers were in a unique position. Despite living on the river and using it as a source of sustenance and income, they were not officially considered a population affected by the dam. Scientists debated whether and how the dam would impact fish populations and the fishers who relied on the river for their livelihoods. Norte Energia, which was responsible for mitigation and compensation programs, was reluctant to make provisions to the fishing community before those debates were settled.[2] Norte Energia also contributed to and influenced these debates through their own commissioned research teams that carried out studies on fish populations and the fishing community during construction, which they were mandated to do. In the eyes of fishers, these studies were too little and too late, offering minimal to no compensation. The fishers felt their lives would be irreparably damaged before anything came of the research. Within the first year of dam construction, as this research was carried out and debates over the impacts of the dam continued among experts on dams and river ecosystems, the fishers began noticing changes to the river and the fish. These fishers, and the many people who lived along the river, also foresaw other changes to come, including displacement from their homes. Even amid the life-altering changes and uncertainty, most of the fishers were not invited to engage with government officials and other groups in decision-making forums, such as CGDEX, the participatory committee described in the previous chapter.

My first boat trip to the Volta Grande, or Big Bend, of the Xingu exposed me to many of the concerns of fishers and those living along the river. Given the Volta Grande's unique natural design, in which the northern-flowing river makes a sudden turn to the south and then back to the north, the distinctive ecosystem is home to dozens of unique fish species and a host of small waterfalls and rapids. The dam would forever change that ecosystem.[3] Water levels upstream of the dam would rise, while the flow of water downstream would be reduced. This would likely alter the way of life for many of the inhabitants of the islands and along the banks of the river. In August 2012, I had yet to see this unique environment that was often at the center of the debates over the dam, so I jumped at the opportunity to travel with Paulo to an island downstream of the dam in the Big Bend.

Paulo was the president of the fishers' union and looked to be around sixty years old. He seemed to carefully consider everything he said, choosing his words in an effort to convince me to trust him and believe in his work. As the union's president, his role was to support its members, who were registered fishers in the region. He processed large amounts of paperwork and seemed to carry out the bureaucratic functions of the job. As president, Paulo served as the official liaison to the state and federal system of fishing unions, which is similar to that of the Brazilian rural workers' unions. Local fishers' unions, such as the one in Altamira, operate at the municipal level. They respond to the state-level organization, which is overseen by a countrywide union organization, the Confederação Nacional de Pescadores e Aquicultores (National Confederation of Fishers). This system of unions, which is designed to defend the rights and interests of fishers, has been federally regulated since the 1970s. The founder of Altamira's union explained that in the early 1990s he began noticing that fishers in other municipalities "had some power because they had a union," so he founded the Altamira chapter in 1997, in order to "fight for the dignity that the fishers deserve."

Paulo wanted to go to the Big Bend partly to show me and another foreigner—an activist—the situation facing the fishers and partly to learn more about the dam's effects on the island where he was born and spent most of his life. It was just over a year since dam construction had begun, and he had not yet traveled down the river in that time. During construction, Paulo and the union sought to grow their reputation and support the fishing community in new ways. As people faced the unfamiliar challenges, the union could help negotiate for better conditions, but Paulo needed to know what was changing and who was supportive of the union. In order to accomplish this, Paulo and a colleague from a closely aligned ornamental-fishing group arranged for a fast motorboat—quite different from the slow, simple motorboats that most fishers use—to take us to the dam site in the Big Bend. I arrived at the port just in time for a long metal boat—which had a metal rooftop, no walls, and was just wide enough for two seats per row—to pick me up. Paulo, the foreigner, two of Paulo's colleagues, and the hired boat captain were on board. As the boat raced away from Altamira, it was a bit loud to talk with anyone on board, but the breeze was a refreshing reprieve from the hot sun and humid air. As we traveled down the river, we passed a series of twisting rapids and small waterfalls surrounded by rocks. Boat captains in this area needed to be well aware of the terrain, as a miscalculation would send a boat into the rocks, which could cause it to capsize. The river is nearly four kilometers wide for much of the Big Bend, and islands are scattered throughout the river, many of which are inhabited.

It took us less than two hours—at least twice as fast as if we were in a simple wooden boat—to reach the construction site of the dam. At the time, the dam blocked three kilometers of the width of the river, leaving a five-hundred-meter passageway open. Paulo asked the captain to slow, so that we could talk about the construction. "This will be completely closed within six months," Paulo stated with a bit of sadness in his voice. "How will boats pass through this area after that?" I asked. Paulo explained that a tractor with a boat trailer would pull smaller boats over land to the other side of the dam. Larger boats, far less common in this part of the river, would be lifted by a large harness that would then be guided by a system of suspended tracks to the other side of the dam. The dam was not particularly high, as its purpose was to block the river and divert much of the water through the manmade canal, but it was nevertheless impressive from an engineering perspective. From the water, the dam looked to be wider than an interstate highway. It was constructed largely out of rock that had been blasted out of the ground. A blue, covered, open-walled tent sat on the end of the dam road, providing workers a shaded respite from the sun. A matching tent was also positioned on the mainland near where the dam would eventually seal off the river. A few yellow buoys marked the distance from the dam that personal boats needed to stay.

The boat continued downstream, passing more lush green islands and through more small waterfalls and swirling waters. We then passed a riverside village that Paulo said was a small community that would be the base of operations of a large-scale, controversial Canadian mining project that had not yet been approved. That village would eventually become the site of intense debates over the dam and the mine. Soon after, we reached the island where Paulo was born in 1953. He lived there until 2003, at which time he moved to Altamira. As we approached the island, I could see about twenty small homes arranged closely together. The roofs looked to be made of natural materials, such as dried leaves and ferns. Paulo pointed out a small school building, which was next to the island's church and a small snack shop. After we climbed off the boat and onto shore, we went into that shop. On the walls were a few calendars, two of which included beautiful pictures of the river and bore the logo and slogan of Norte Energia: "Development with respect for the people and the environment." The shop was the front room of the home of the shopkeeper, Edilson. The wooden house with a tin roof stretched back through three other rooms until the kitchen, which was where we ate lunch.

During the two hours before we ate, we spoke with Edilson and Dilermando, a fisher. Edilson was a larger man who wore a plain purple T-shirt that was a bit ripped. His facial hair was not a full beard but instead looked as though he had not shaved in the previous few days. Edilson seemed forlorn and he did not even try to force a smile during our conversation, but he was gracious and welcoming, more than happy to talk with us about the challenges he and other residents were facing due to changes to the river and the dam's construction. He started by lamenting the transportation system to get around the dam. Norte Energia officials had recently held a meeting, which he had learned about through a friend, to explain the system, so it was fresh on Edilson's mind. "What if there is an emergency?" he asked. "We will have to wait twenty minutes to get through!" He was also upset with Norte Energia for holding the meeting in Altamira, rather than on the island, and failing to invite the residents of the island, who would be some of the people most impacted by the transportation changes.

As Edilson finished explaining the future challenges posed by the dam, Dilermando stepped into the doorway of the shop. He was about fifteen years younger than Edilson and much taller and skinnier as well. Dilermando was wearing a shirt and hat that matched, both with the phrase "Indigenous Communication Program—Belo Monte." He did not look Indigenous and was unlikely to be working closely with officials at Norte Energia, as he began complaining about a number of things, particularly the reductions in the fish population and the lack of response from Norte Energia. "Last year at this time you could see fish jumping out of the water right here," he said, pointing to the water just a few steps away. "I recently went fishing near here and caught six or seven kilos of fish. Last year at the same time it was over twenty!" Neither Edilson nor Dilermando knew for certain why fish populations were declining, but they observed that the river downstream of the dam was dirtier and had lower water flow since construction had begun.

Dilermando said that Ibama, the federal organization that oversees social and environmental protections during development projects and issues construction and operating licenses to companies like Norte Energia, made a recent visit to the island to determine if there were any problems. After a few days of investigating, the officials, Paulo remarked, simply told the residents they had nothing to worry about. These residents, like many others who lived and worked along the river, were becoming increasingly concerned with their situation and dissatisfied by the support that they expected but were not receiving from Norte Energia and government officials. In addition to receiving no compensation for the effects the dam was having on their lives, they also felt as though they were receiving little information about what was happening. In the months and years following this first trip, I would return to this island and to some of the other riverine communities we had passed with the public defender and anti-dam activists. I would see the physical changes to the landscape and the shifting ways the communities spoke about the impacts of the dam on their lives.

In the two months following that first trip to the island, I learned a great deal more from the fishers about their situation. The areas near the construction sites were particularly vulnerable, due to planned explosions to break up the rocks and due to the chemicals that were used in those explosions. The explosions, the fishers argued, confused the fish, making some abandon areas in which they usually spawn and live and causing others to die. The chemicals used around the water, the fishers suggested, were also harming and killing the fish. One woman who fished for a living told me that, for the first time, she began to see bloated, discolored fish floating near dam construction sites.

Fishers also had unspoken territories, and they argued that people who were impacted by the dam's construction would move upstream or downstream into areas where other people already fished. This would cause conflicts and speed up declines in fish population. Furthermore, some fish species rely on the changing levels of the river. On my trip with Paulo, he and his colleagues explained that some fish species spawn in areas that are protected during the season when water levels are low. In those places, the young fish are sheltered from predators. The dam would decrease the changes in water levels, which would threaten some of these species.[4]

In addition to these changes to the ecosystem, the coming displacement from homes and places of work deeply troubled many of the fishers. Most people who fished for a living had homes and property on the banks and islands of the river. They stayed there the majority of the time, but the urban area of Altamira served important purposes for the fishers. They sold their fish in the city, had other homes along the shores and streams in the city, and often had family who attended school or were otherwise based in the city. Rodrigo, for example, had a typical routine for a fisher. Each week, he would spend five days fishing on the river. On the sixth day, he would make his way back to the city, sell his catch to a wholesale fishery positioned on the banks of the river, and spend another day with family and friends.

Born on the banks of the Xingu in the Big Bend, Rodrigo had been fishing his entire life. His father was first a rubber tapper and then turned to fishing to feed the family and make a small amount of money. Rodrigo never learned to read or write, signing his name with an X as he grew older. As dam construction began, Rodrigo was in his early fifties. He was dark-skinned, strong, and extremely friendly, even as he faced some of the most challenging living conditions. He had not participated in much activism but was outspoken enough that other fishers looked to him for guidance.

Rodrigo's urban home was typical of that of most fishers in the city. Not far from the banks of the Xingu, he lived in a neighborhood of *palafitas*, hand-built stilt homes elevated to avoid flooding in the rainy season. When the river levels were high, water flowed under most of the houses. In the dry season, the dirt areas under and near the houses were gathering places but were also filled with trash. The lack of a proper sewer system meant that dirty water and toilet waste often collected under the homes, directly entering the streams and river when water levels were high. Their stilt homes were at risk of falling over. People lived in cramped conditions and they experienced health problems due to polluted water and mosquito-borne illnesses. Rickety planks of wood served as bridges between homes and to the streets of the city. One of the arguments that proponents of the dam made over and over again was that these were inhumane and subhuman living conditions. On my way back from a personal tour of the dam site, for example, a representative of Norte Energia who had taken me to the dam

FIGURE 5.1. Prior to displacement, many fishers lived along the river in homes like those pictured. (Photo by author.)

made a point to drive through the *palafita* neighborhoods, saying "Look at how these people live in such bad conditions!" He explained how Norte Energia would build new homes in which everyone could live with dignity. The proponents of the project who cited these issues had a strong argument. Indeed, there were plenty of people who lived in these neighborhoods who looked forward to new homes in new neighborhoods. But for fishers like Rodrigo, the prospect of moving was more complicated. Their livelihoods were tied to the river, even when they were in the city.

The union and some of the fishers pushed to acquire land that was slightly outside the center of the city yet still on the banks of the river. They could have both homes and businesses there, which would allow them easy access as they came and went. This became an incredibly slow process and one that was not completed before my last trip to Altamira at the end of 2017. Instead, many fishers were relocated to urban resettlement communities far from the river, just as they had feared. Rodrigo, for example, explained to me in 2017 that he used to live fifty meters from the river but was now five kilometers away from it. It would cost him the equivalent of about fifty U.S. dollars to get himself and his boat equipment to the river just once, which he could not afford.

The people who fished for income lived in precarious conditions and had few avenues through which to make claims. They faced an impending loss of easy access to the river that they depended on and were often left out of participatory spaces of deliberation and decision-making. For example, other than the

FIGURE 5.2. Urban Resettlement Community (RUC). Relocation programs moved many fishers to RUCs far from the river and city center. (Photo by author.)

ornamental-fishing group, whose membership was relatively small, fishers were not represented at CGDEX. The local fishers' union did not have a position and rarely attended meetings with government officials or representatives of Norte Energia. Furthermore, some of the fishers were in a strained relationship with the union, as not everyone thought Paulo would work hard enough to defend their rights and support their needs. At the federal level, the Ministério da Pesca e Aquicultura (MPA—Ministry of Fisheries and Aquaculture) had not originally been invited to be a member of CGDEX or even participate in the technical groups. Even as the agency that oversaw and was meant to serve the country's fishing community, the ministry had little influence over Belo Monte and few avenues through which to support those who fished for a living. Many of the fishers were left without an agency or individual to mediate their claims and demands.

Given the lack of attention to the fishers' precarious situation and their lack of recourse, it would be up to them to show that the dam was impacting their lives in detrimental ways and that the plans for compensation and relocation were insufficient. They would, over time, have to counter the narratives of Norte Energia and other proponents of the project that suggested the fishing community would benefit from the dam's construction. This would mean creating a group of trusted allies who could serve as mediators of their claims and help them devise strategies to make the government and Norte Energia listen to their demands. Their first large act of resistance would accomplish some of this to surprising effect, mobilizing state actors, providing an education in civic engagement, and giving these humble fishers the confidence to persist through years of barriers and challenges.

ACTIVATING THE STATE

Some fishers were involved in incidental protests in Altamira just as dam construction was beginning, but they were not well organized.[5] A year later, they mobilized in a way that would lead to significant impacts. Under the hot sun of a September afternoon in 2012, a small group of about a dozen people who fished for a living, supported by the fishers' union, Xingu Vivo, and other regional nongovernmental organizations, gathered on the banks of the river in Altamira. By that point, the fishers had begun noticing changes to the river. Fish populations were diminishing, construction was about to cut off easy navigation on the water, and they knew that more serious changes, such as displacement from their homes and further disruptions to the river, were looming. They had also noticed that resources and opportunities were coming to the region, but they felt left out and unable to access them. In addition, the fishers recognized that there was a great deal of attention on Belo Monte from the federal and state government, as well as nongovernmental organizations, so it was an ideal moment to make a statement. They wanted to call attention to their situation and hoped to receive some kind of compensation for their suffering. More importantly, they wanted to do *something*. While most of them were not opposed to the dam per se, they felt silenced.

This small group of fishers wanted to break that silence with an overt, direct act of resistance. Given their difficult situation and the opportunities of the moment, the group decided they would fish in the newly prohibited zone near the dam construction site in the Big Bend. As the fishers prepared their small wooden boats for their trip down the river, local reporters arrived on that sandy shore to document the first day of what would turn out to be a long protest. Rodrigo, the most outspoken of the fishers, was an important organizer of the protest. He spoke to the reporters about their motivations for their act of resistance:

> We need this area [of the river] ... We now see a company, Norte Energia, preventing us from working in this area. I've been a professional in this area for twenty-eight years, and really we aren't against [Belo Monte]. We want them to understand that we need our rights as fishers of the Xingu River ... We are using lawyers that are representing the union to request our rightful compensation that we have as professional fishers. And until this moment Norte Energia hasn't given us anything. They have abandoned us and they are trying to do what was done in

Tucuruí... We don't accept this... Norte Energia will make a mountain of money off this dam and the fishers will be left with what? We need the directors of Norte Energia to actually discuss the situation of the fishers with us!

Rodrigo was suggesting that the union was not making progress through legal mechanisms, so they would try another approach to gain leverage. Along with the others who were engaging in the protest, he hoped that the protest would draw more attention to their plight, forcing Norte Energia to listen and respond to their demands for compensation and their claims that the river was changing. Paulo, the leader of the fishers' union, shared similar thoughts with the press: "We are not against the dam, but we are fighting for our rights, for our livelihoods, and the respect we deserve." Paulo later suggested to me that he was supporting the protest because it would provide more leverage as he met with officials from Norte Energia, who would not be able to discredit their situation after such a public act.

As the interviews with the press concluded and after a sudden deluge of rain, we quickly put our gear under tarps, as the boats were otherwise uncovered. Alberto, the fisher whom most people credited with the idea of fishing in the prohibited zone and who would become the de facto leader during the monthlong protest, managed the logistics. He had a matter-of-fact style, saying who would do what and when they should do it, and he did not shy away from confronting people with whom he disagreed. Alberto was the son of a Kayapó Indigenous person, and he often used the tribe's traditional black body paint to represent himself as such. Others often called on him for inspirational words, and, as described in the prologue, Alberto represented the protesters at the conciliatory hearing at the end of the protest and occupation.

Alberto directed me to step into the boat of a married couple whom I had yet to meet. We were the first to leave the shore, and as we puttered away under the power of a small outboard motor, I looked back toward the city. The view was striking. The fishers had affixed small Brazilian flags to the front of their boats before setting off for the four-hour journey. Over a dozen small, wooden boats, each with a Brazilian flag fluttering in the wind, were strung out in a line as Altamira slowly disappeared behind us. These flags may have been a practical attempt to avoid violent repression by the police, but the fishers were also making a statement that this was their country, too. While the government and others in power were intent on building this dam, the fishers would have a say in how this was done. As Paulo said in an enthusiastic speech on the first day of the protest, "Every one of us here has rights. Let's fight for our rights! It is only like this that we will show that we are Brazilians and that our rights have to be secured!"

On the first night after we left the city, about ten fishers, an activist from Xingu Vivo, another researcher, a journalist, and I slept in hammocks on a fisher's property that was a short boat trip from the dam site. The actual protest at the dam site would begin the next day. The journey had taken over four hours, and we arrived after sunset. By that time, the mosquitos were thick and swarming, so the four of us who were not fishers, and thus not accustomed to the bugs, quickly covered ourselves in repellent. We hung our hammocks, made a campfire to cook the meat the fishers had brought, and talked about what might happen in the coming days. Nobody knew what exactly the group planned to do or what challenges they would face as they began fishing in the prohibited areas. Alberto, for example, commented, "We don't know what we will encounter and come up against as we go down the river. The police, resistance from the company, or other things might challenge us." The possibility of arrest and other difficulties did not deter them, but they all seemed unclear as to what they would actually do the next day. The only thing that was clear was that nobody there knew exactly what they were doing. As I spoke with the fishers, I realized they had little idea of their specific objectives and how and to whom they would make their demands.

The next day, we awoke as the sun came up, packed our things, and after more discussions about facing the unknown challenges that awaited, we headed to an island in the prohibited zone. There, other fishers began arriving, and the group swelled to about thirty people in over fifteen boats. As some of the fishers waited to set off in their boats, a larger, fancier boat, which everyone presumed to be from Norte Energia, passed back and forth, slowly coming closer to the island. After it passed by about a hundred meters from the shore, the boat left the area. The fishers were clearly drawing some attention.

The group would go on to fish in these prohibited zones and camp on this and other islands for nearly a month. Despite the isolation, they developed what amounted to a small village. They set up separate covered areas for sleeping and eating and ran generators to power lights and cooking equipment. A few women cooked meals for everyone who was there, and others helped themselves to the food before heading out on their boats or working on something for their encampment. Each night, the group talked about their strategies and plans moving forward.

As the protest went on, it became clear that many of these fishers were motivated by the prospect of people in positions of power hearing about their hardships and then responding. But they were entirely new to activism and the act of protest. They were inspired to make a claim for justice and had received some support from established institutions but did not have a sense of what to expect. The fishers had to decide whom to target for what demands and be flexible in waiting for a response from an unknown entity. For those engaging in protests, the multiplicity of targets had the potential to confuse and weaken their efforts. The fishers knew they needed Norte Energia and municipal, state, and federal government officials to each answer to different aspects of their needs, but they did not know who could provide what. They framed their demands in ways that

FIGURE 5.3. Fishers traveling to prohibited zone on the first day of their protest in September 2012. (Photo by author.)

were broad enough to target a variety of actors, yet specific enough to help build a network of supporters.[6]

Given the uncertainty and the complex nature of the situation, the fishers looked to everyone for advice, even as they tried to shape their engagement in their own ways. They wanted to guide their own futures but sensed the benefits of being guided through the mechanisms of how to do so. In the short term, this left them somewhat confused as to whom they could trust and what each allied group and individual wanted from the protest. Did the union have their best interests in mind? Were the social movement activists aware of their needs? What were the best steps to move forward with their demands? As they sorted through these types of questions and continued to engage in their resistance, these humble fishers became engaged citizens, activists, and persistent claims makers. In the long term, they learned whom to trust, built the skills and knowledge to make demands on their own, and developed a strong voice for themselves. They built a constantly evolving network of people and organizations they felt were truly acting for their benefit.

At the encampments, the fishers continued to temporarily live and fish in the prohibited areas. Back in Altamira, the protest was generating a great deal of public interest and mobilizing government actors. I traveled to and from the protest sites throughout the month, which allowed me to witness how government agencies, Norte Energia, and other actors responded to the protest. After just a week of the protest, it was clear the fishers had succeeded in drawing the federal government's attention. While a small group of committed protesters continued their act of resistance, a high-level representative from the Ministry of Fisheries and Aquaculture (MPA), Fernando, met with over a hundred fishers at the union headquarters. He explained how the MPA would work with the union and other local groups to ensure that Norte Energia would hear the fishers' demands. Fernando immediately arranged a second meeting with representatives from each of the groups who attended the first meeting, including the union, two other associations that represent different sectors of the fisheries, the Movimento dos Atingidos por Barragems (MAB—Movement of People Affected by Dams), and officials from Casa de Governo (Government House). The Casa de Governo was the "local representative of the federal government" that was designed to "accompany the demands of the region and find solutions together with federal and municipal organs and together with the business—the Norte Energia consortium—guaranteeing the presence of the State locally [in Altamira]."[7] At the beginning of this second meeting, Fernando explained his approach to addressing the protestors' concerns: "I haven't yet requested a meeting with [various federal level governmental agencies] and Norte Energia to bring the demands of the fishers forward. I first wanted to hear from you . . . make a quick evaluation to understand the process . . . and determine how we can improve the deliberation process that was originally created by the fishers."

Throughout the meeting, he and the federal representatives from the Government House were interested in discussing the concerns of the local community and putting pressure on Norte Energia in the name of the fishers. As the participants explained their numerous concerns with the changing river, the threats to their livelihoods, the difficulties with their equipment, and the failure of Norte Energia to respond to complaints or offer compensation to those being affected, Fernando and the other federal representatives asked clarification questions and provided opinions on what could be realistically expected. As the group developed ideas of how to improve the situation, many stressed the importance of Norte Energia's presence at meetings and the mediating role that the federal government could and should play. By the end of the meeting, they had drafted a document that summarized their ideas for moving forward, primary among them being the creation of a committee run by the federal government that would be dedicated to the fishers. The way these government officials responded to the protest highlights that the fishers had mobilized the MPA and other state actors.

This mobilization of government institutions like the MPA was important but also limited, showing the constraints of direct acts of resistance in the face of large infrastructure projects controlled by private entities. Targets of protest were either the state, which was restricted in its ability to actually make the changes that were being demanded, or Norte Energia, against which protestors had limited leverage due to the consortium's lack of accountability to the people. Rather than having the power to directly meet the demands of protestors, state agencies like the MPA had to use their position to force Norte Energia to abide by laws and regulations designed to meet the needs of affected populations. This was further complicated by the fact that government officials at all levels (municipal, state, and federal) were still responsible for providing ongoing services and infrastructure to the population. In this type of situation, it was possible for government agencies at all levels to knowingly or unknowingly shirk their responsibilities, claiming that they were either not responsible or that they were doing everything they could to deal with the situation. The avoidance of responsibilities in this context can be seen as another example of what Shalini Randeria calls the "cunning state."[8] The MPA, which appeared to be trying its best to support the fishers, seemed to understand this dynamic and was thus careful to listen to the difficulties that these fishers faced, determine which governmental agencies at what levels needed to be involved in future decision-making processes, and decide how these government actors could best pressure Norte Energia to meet its obligations. The new relationship between the fishers and the MPA, however limited, was valuable.

The protest also strengthened the existing connection between the fishers and the Public Defenders' Office of the State of Pará, represented by Andreia Barreto, the public defender who had committed the most effort to cases surrounding Belo Monte. Earlier in 2012, prior to the protest and occupation, the public defenders' office, led by Andreia, had started building relations with the fishers and riverine people. She traveled to communities along the river, in order to hear from the people about their experiences and offer her services as a public defender of their rights. Through these efforts, she had started to form relationships with some of the fishers and leaders of those communities. The protest and occupation crystallized these connections and further increased the office's institutional legitimacy in the eyes of the fishers and other dam-affected residents. Even though the union had begun working with a private lawyer, the fishers who had originally organized the protest chose to work with the public defenders. Some of the fishers and the social movement activists that accompanied them were concerned that the private lawyer was more interested in filling his own pockets than actually securing long-term rights and compensation for the fishers. Others simply trusted Andreia because they knew her and the work she had been doing over the previous year. At the end of the protest, when the fishers were joined by Indigenous groups and occupied the dam site for ten days, Andreia and others at the public defenders' office became the primary legal support for the group. They visited the occupation site, spoke with the fishers about their needs, and helped them generate a list of demands.

The fishers' protest and occupation concluded with the two-day conciliatory hearing at the dam site described in the prologue. Alberto and one other fisher, along with Andreia and a small team from the public defenders' office, represented the group of fishers who had been protesting for nearly a month. They pushed for material compensation and access to deliberative spaces through which government officials and Norte Energia representatives would hear their claims. After two days of negotiating with the groups who had occupied the dam site, which included Indigenous groups, boat captains, and the fishers, Norte Energia agreed to conduct more studies *with* the fishers. The facilitators of the hearing argued that this collaboration would allow the fishers to show researchers the realities of life on the river. Norte Energia also agreed to invite the fishers to one of the official, federally mandated "social accompaniment forums," a space coordinated by Norte Energia that was designed to provide an avenue through which communities could hear about and discuss the progress of Belo Monte and its effects on the region.

Andreia and some of the social movement actors that supported the protest were not pleased with these outcomes of the conciliatory hearing. They felt as though the fishers, who had committed themselves to the protest and were suffering due to dam construction, had not gained anything. At the time, I shared this assessment. After toiling for so long, the fishers were only guaranteed more opportunities to discuss their situation with those in power. The promise of deliberation and negotiation had quelled the protest, just as the demonstrators had gained leverage. From the perspective of us observers, the potential for achieving material gains, such as compensation for losses, seemed remote.

While Andreia and others seemed forlorn, many of the fishers were relieved and satisfied with what they had accomplished, at least for the moment. The possibility of actually engaging in ongoing, direct negotiations and decision-making processes appealed to many of them. Norte Energia had previously restricted access to the forums to a limited number of officially sanctioned civil society groups and municipal representatives, so the fishers felt that securing an invitation was a real success. In the end, these opportunities turned out to be the beginning of a long road to obtaining some gains and compensation.

FROM DISSENT TO DELIBERATION

By gaining access to these opportunities and mobilizing state actors to work on their behalf, the fishers, through their protest, generated long-lasting consequences. They strengthened parts of government, built a network of advocates, and set themselves on a course to be engaged residents who would have a say in their own future. The fishers did not protest again, although most of them said they would be willing to go back to the dam site if necessary. Instead, the months following the protest were marked by direct negotiations between the fishers' union, government officials, and Norte Energia. One municipal official who worked closely with the fishers explained the change: "Now [after the protest] is a moment for dialogue. The fishers are trying to organize better now than in the past... they have been making connections with people within the [municipal and federal] government to try and get what they want." Their protest had helped the fishers create new avenues of negotiation and find new advocates to mediate and translate their demands.

At the first Norte Energia forum meeting to which the fishers were invited, the MPA initiated the Commission of Fisheries and Aquaculture, the steering committee that Fernando discussed during the protest, to address specific concerns related to fisheries. In addition to the MPA, the commission was comprised of federal agencies that represented environmental and Indigenous concerns, the state-level secretary of fisheries, municipal governments from the region, the public defender of the state of Pará, and the fishing sector, including the union and other civil society associations that represented various types of fishers. Over time, the Altamira-based civil society groups began to focus on the housing situation for fishers, and the commission became the principal avenue through which they could voice these concerns. The commission allowed for direct negotiations between state agencies, Norte Energia, and the fishers in developing a plan of action prior to the flooding of many fishers' homes. This committee would also be a place to discuss what to do about the commercial businesses that supported the fishers' work in Altamira, most of which would also be flooded.

Julia, a representative of the MPA, participated in nearly all of the meetings in Altamira and worked closely with many of the groups representing the region's fisheries. Nearly a year after the protest, she confirmed that the fishers' use of direct action influenced the creation of the special Commission of Fishing: "If it were not for the protest, the commission would not have happened. That [protest] put the pressure on Norte Energia. The fishers called for the ministry, so Norte Energia had to respond." She added, "This was important for the ministry as well, because [other sectors of the] government started to recognize us."

Julia's statements are telling, because, as is evident throughout the story, the act of direct resistance led to multiple opportunities for fishers to participate in negotiations with various sectors of the government and Norte Energia. The union wanted the support of the federal government and saw the opportunity for partnerships, and the protest opened avenues for these partnerships to build. Additionally, Julia's follow-up statement reflected some of the frustration felt within the ministry over not being included in other processes related to Belo Monte. The protest brought attention to the fishers' situation, and when the union called on the MPA, this forced not only Norte Energia but also other parts of the government apparatus to recognize the work of the ministry.

Through the special commission and more active participation in other spaces such as CGDEX, other institutions and agencies formed new alliances as well. Julia, for example, found she had common goals with Mari, from the

Government House in Altamira, and Andreia, the public defender. They connected with one another through their shared support of the fishers and began collaborating to defend the fishers moving forward. They also worked together as a unified voice on other issues. At another social accompaniment forum on general housing issues, for example, Mari and Andreia together pressured Norte Energia officials to provide clear information on the urban resettlement process. The three women became such close partners that Andreia referred to them as the "bando das mulheres," or gang of women. Andreia was clearly proud that they were three women using their institutional authority to stand up against a powerful entity. Their work together also strengthened their own institutions and their ability to make an impact with Norte Energia and other government agencies.

The fishers' support system was growing through these connections. Some of these state institutions, such as the public defenders' office and the MPA, began serving as mediators of the demands that the fishers wanted to make on Norte Energia. Julia and Andreia, specifically, began translating demands by reframing the fishers' claims to pressure officials would to act. For example, Julia explained that the MPA helped to "orient Norte Energia," holding meetings with technical teams from Ibama and the construction consortium. In these meetings, they talked about what the fishers wanted and what Norte Energia was obligated to provide. In communicating what the fishers were requesting, she used language such as "according to the law" in order to pressure Norte Energia. Her translation served not only to ensure that the fishers understood the technical language and processes of Norte Energia but also to verify that Norte Energia clearly understood the demands of the protestors and that they had a responsibility to respond to those demands.[9]

Julia and Andreia performed similar roles in other settings. During the meetings of the special commission on fisheries, the group—which was comprised of representatives from multiple government agencies, Norte Energia, and civil society groups—discussed a particularly contentious housing settlement for the fishers whose homes and businesses would be flooded. Members of the union and other fishers expressed their concerns. One fish shop owner conveyed her frustration: "Norte Energia does the best for itself, picks the cheapest option. The population of Altamira did not ask for Belo Monte. We are happy in our houses even though Norte Energia says they're poor quality... You need to think more and respect the population... I'm not against Belo Monte. I'm against the pain you have brought to us. You don't listen to us, but this will not go forward because we will go to the streets and get what we deserve."

Others shared similar thoughts, pleaded with Norte Energia to understand their situation, and threatened to disrupt construction again through protest. Julia then stood to speak, bringing the conversation back to concrete requests they had made that, if fulfilled, could allow the fishers to play a more significant role in the deliberation process. "We are trying to facilitate an agreement and solution," Julia said. "The ministry has asked for the potential sites to be visited with the fishers, and until now I do not know if that has happened. Second, we asked that the documents [regarding the housing options] be given to the fishers [by Norte Energia] ... Neither I nor the fishers have a say in the selection of a site ... and the lack of technical information from Norte Energia is causing huge problems. The fishers have been very clear about their concerns and desires, and Norte Energia has not addressed them."

This anecdote shows how an advocate could help mediate discussions and translate fishers' demands in ways that could have impact. Through a network of advocates and the use of protest, the fishers pressured government officials to follow through on promises to mitigate negative impacts of construction and include dam-affected people in decision-making processes. Through their claims-making efforts, the fishers were also building state capacity and strengthening state agencies and institutions, like the MPA and the public defenders' office. Other accounts have suggested that civil society can bolster government institutions and programs because of the absence of a strong and capable state. The fishers, in contrast, found success in mobilizing government agencies not because of a weak state, but in part due to the presence of a private entity, Norte Energia.[10] The construction consortium became the target of claims making. The fishers—along with the advocates they had mobilized—were beginning to find ways to leverage Norte Energia into meeting their demands, even if negotiation was incredibly slow.

FIGHTING FOR RECOGNITION AND CHANGING COLLECTIVE IDENTITIES

On the one hand, the fishers' successes during the first few years after their protest were impressive. They had drawn attention to their plight, mobilized and empowered officials of the state to work on their behalf, and gained access to spaces of deliberation. On the other hand, the group of fishers had not made many material gains. In addition to losing their rural fishing homes along the banks of the river, most of them were also displaced from their urban neighborhoods, which had given them river access. They were not given financial compensation for changes to the river that were impacting their livelihoods.

In addition, the people and institutions who had been helping mediate their demands were also changing. The office of the state-level public defender closed in 2015, and as a result, Andreia, the public defender who had stood with the fishers during the occupation, moved to another city until the Altamira office was reopened at the end of 2016. Also in 2015, the federal government closed the MPA and incorporated it into the Ministry of Agriculture. The fishers had helped empower the MPA to work in Altamira, and the ministry had been a key connection with the federal government. Its closure weakened the fishers' voice at the national level. Furthermore, the small group of fishers who had inspired the protest and occupation were losing confidence in the union that represented them. That union had stood alongside them during the protest, but some of the fishers believed the leadership of the organization was only working for themselves, rather than all of the fishers and those most directly impacted by the dam.

After multiple agencies closed and the union's support wavered, the fishers' struggle could have lost momentum. Instead, they continued to make progress. The fishers were persistent and they had built a significant network. The new commission to discuss issues related to fishing endured and the fishers continued to attend meetings. Importantly, other institutions became key supporters and mediators of claims, taking up the fishers' cause and helping to push forward new demands. The public prosecutors' office, and particularly Thais, the prosecutor who had attended the commemorative event with me, became an indispensable advocate for the fishers. The Socio-environmental Institute (ISA) also intensified their work with the group. Xingu Vivo continued to provide support, and the federal public defenders' offices joined the cause when they finally opened their office in Altamira in 2015.

As this new support network grew, the struggle shifted focus and the fishers transformed their collective identity from fishers to riverine people. Collective identities refer to the categorizations that cognitively, culturally, or emotionally link individuals to groups and that emerge from related processes of external and internal identification.[11] Rather than make demands based on what the group did for work and livelihood, which was fish, they began making claims based on where they lived, along the banks of the river. This important change happened for various reasons. They had to unify in a way that helped others see them as a dam-affected population with specific needs, and the forced displacement of their group seemed quite clear to outsiders. By 2016, most of the urban and rural resettlement process was completed, and thousands of fishers and riverine people found themselves living in urban resettlement communities far from the river's edge. They felt that Norte Energia and the government carried out a string of unjust actions during the forced displacement process.

Their struggle for recognition—as riverine people impacted by the dam and whose rights were violated in the resettlement process—was a lengthy and arduous process. Norte Energia had long downplayed the effects of Belo Monte, using their own contracted researchers to make their claims. The riverine people, along with nongovernmental organizations, researchers, and legal officials, countered that narrative by using their own research and evidence-based reasoning. They made advances by shifting predominant understandings of the impacts of dam construction.[12]

ISA was one of the key groups to help the fishers take back the narrative with research-based evidence. The group had worked closely with traditional communities in the region and so was a natural ally. Ana de Francesco, a doctoral student from a university in the state of São Paulo, UNICAMP, teamed up with ISA to carry out her research and help find ways to support the communities. Her research led to ISA's publication of the *Atlas dos Impactos da UHE Belo Monte sobre a Pesca* in 2015. This glossy, fifty-five-page booklet summarizes the impacts that the dam was having on fishing from the perspective of the fishers. Construction had started four years previously, but up to that point, few publications testified to the perspective of the fishers. As the booklet's introduction states, "The traditional fishers from the area affected by the Belo Monte dam are unanimous in their observations of the negative changes caused by the installation of the dam, but they have never been seriously heard as the leading experts of their own territory."[13] The booklet sets out to "give voice to these fishers" and share their knowledge. The book includes dozens of beautiful pictures and descriptions of the river, the people who fish and live along the river, and the dam construction. In addition, it includes maps, graphs, and charts that intend to convey the extent to which dam construction has impacted the region. The booklet also contains many quotes from the fishers about their lives, often positioned next to a picture of the person's face. Through these descriptions, the reader comes to know the history of the region and the impacts of the dam's construction. The booklet was one of the first research documents to challenge Norte Energia's narrative; it set the stage for claims-making efforts, discussions, and further research on riverine communities and fishers.

Over the subsequent year, Thais from the public prosecutors' office and Ana, along with others from ISA, followed up on the publication with other events that helped give voice to the fishers' plight. In June 2015, Thais organized an "interinstitutional inspection" of the areas along the river that were impacted by the forced removal of people. During that inspection, the groups verified a lack of compliance with the licensed conditions of dam construction. In April of the following year, 2016, Ana organized an event called Riverine People's Dialogues, in an effort to portray their lives. Alberto, Rodrigo, Jose, Gloria, and the other fishers who had begun the fight with their protest four years earlier remained engaged in these events, and the formal participatory meetings continued. The narrative was changing, but more rigorous and timely research about the displacement process was needed to make material advances. The anecdotal stories of the people with whom Thais worked suggested that Norte Energia was carrying out the relocation process based on inaccurate information and in ways that violated human rights laws. Through the booklet, various events, and participatory meetings, the fishers and riverine people had begun to contest Norte Energia's narratives with their own stories, but dam-affected populations lacked the credibility for their claims to be taken seriously.

The tension and disagreements between the riverine people and Norte Energia were the latest reflection of the long-standing debate over knowledge, experience, and science in relation to Belo Monte. In 1988, a pro-Indigenous group in

São Paulo published a book titled *As Hidrelétricas do Xingu e os Povos Indígenas* (*The Dams of the Xingu and the Indigenous People*), in an effort to show the negative impacts that the proposed dams on the Xingu would have on Indigenous groups.[14] In 2005, a group of researchers, journalists, and activists put together a three-hundred-page book to "deepen the debate about the proposed construction of the Hydroelectric Complex of the Xingu," a direct objection to the information being distributed by the proponents of the project.[15] Finally, in 2009, the Panel of Specialists, a group of researchers from universities around Brazil, published their *Critical Analysis of the Environmental Impact Assessment of the Belo Monte Hydroelectric Project*. The twenty-two-chapter document directly addressed the problems and gaps in the official environmental impact assessment, which would guide the mitigation and compensation program. In the report, as in those that came before it, the researchers address the social, economic, and cultural impacts of the dam, as well as how Indigenous populations, the natural environment, and residents in the urban and rural regions near the dam would be affected in ways that official documents had disregarded.[16]

After construction of Belo Monte began, the debates over knowledge intensified. Thais, ISA, and the riverine communities realized they would need to find new ways to show the experiences of the people who had lived along the river. Thais, in particular, felt the need to build a stronger case that was backed up by rigorous research to counter Norte Energia, which supported its narratives with its own scientific research. Many public prosecutors in Brazil employ a variety of techniques and partner with various organizations and researchers in order to strengthen their arguments.[17] In this case, Thais wanted to enroll a team of experts to carry out research about the riverine people of the Xingu and the changing river conditions that would impact those who relied on the river for their livelihoods. She hoped that a citizen-science alliance—to extend sociologist Phil Brown's concept—would strengthen the anecdotal authority of the people impacted by the dam.[18] In June 2016, Thais traveled to São Paulo to ask the president of the Sociedade Brasileira para o Progresso da Ciência (SBPC—Brazilian Society for the Advancement of Science) for support. Thais quickly received the assistance she needed.

The research would have to be done fast by serious scholars. With the help of the SBPC, Thais sent out a call for researchers and received a great deal of interest. Less than a month after meeting with the president of the SBPC, Thais and thirty-one researchers from universities and organizations around Brazil met in São Paulo to discuss the details of the research that needed to be done. Thais had just met with Ibama, the Brazilian licensing agency, to discuss her plan, and they agreed that they would hold a public meeting in November to share the results of the research. This gave the team of researchers just over three months to carry out their work in and around Altamira and produce reports. But Thais and the researchers were focused. Those who organized and participated in the project hoped this focus would provide the precise information on which specific claims for compensation could be made.

On November 11, 2016, the public meeting to share the results of the research took place. Over eight hundred riverine people, far in excess of the two hundred people that the organizers had planned to host, attended the meeting. Nearly a dozen government agencies were also present. The researchers, along with some of the residents who had assisted with the research, presented the findings of seventeen different reports, which were later published as a book, *A Expulsão de Ribeirinhos em Belo Monte* (*The Expulsion of Riverine People in Belo Monte*).[19] The book details how the state and Norte Energia dispossessed riverine communities from their lands, highlights the inadequate efforts to mitigate the negative impacts of dam construction on those living along the river, and provides recommendations to compensate and support those communities moving forward.

A month after the meeting, the Riverine People's Council formed, with twenty-four official members representing the riverine communities affected by Belo Monte. Shortly thereafter, in early 2017, Norte Energia granted the council four seats in a newly created Working Group of Fishing, which was a small group of officials from the construction consortium and key representatives of the fishing community dedicated to addressing the pressing needs of riverine communities and people who fished for a living. These seats represented the first formal recognition of the group by Norte Energia, so it was a significant accomplishment.

Less than a year later, I was at Gloria's house celebrating her birthday and the accomplishments the council had made, about which the fishers, the public prosecutor, and others who had helped along the way told me with great pride. The council had taken it upon themselves to identify all of the riverine people impacted by the dam, a significant undertaking but one they could control. Additionally, some families were finally seeing material compensation. The council had managed to receive guarantees from Norte Energia for monthly payments to the families of riverine communities, and some people, like Gloria, had already been relocated to land along the river.

These successes were accompanied by less obvious but still consequential moments of progress. For example, the day before the celebration at Gloria's house, representatives of Ibama and Norte Energia referenced the Riverine People's Council by name during a meeting of the Working Group of Fishing. Given what everyone had told me about the council's accomplishments and the festivities planned for the celebration, I was not surprised by this. After the meeting, I began to realize that the comments of those officials represented a significant step forward. As the meeting concluded, a few people invited me to lunch, which would be served at the small outdoor cafeteria on the university campus. Sergio, the leader of the ornamental-fishing group who also participated in CGDEX, gave Rodrigo, Jose, and me a ride to lunch in his truck. In the truck, the fishers were visibly excited as they talked about how Norte Energia and government officials were referencing the council for the first time. For these fishers, when officials in positions of power mentioned the council by name, they were formally recognizing the riverine people as an affected group. This recognition was a significant step forward in their claims making.

The fishers and riverine people embarked upon their long struggle with the goal of achieving compensation for the impacts that dam construction had on them, but they did so by fighting to be seen and recognized. In order to gain that recognition, the community had to generate new, legitimate knowledge, which they were able to do by mobilizing actors both within and outside of the state apparatus, including social movements, unions, nongovernmental organizations, government ministries, and the legal institutions.

While all of these actors played important roles in the struggle for justice for the people who lived and fished along the Xingu River, the public defenders' and public prosecutors' offices shaped the debate and were integral to making gains for these groups. How did these legal agencies, which operate between the boundaries of state and civil society, become so important and effective? How did legal officials build and manage relations with dam-affected groups and others both within and outside of the government? What can we learn about the role these agencies play in the reconfiguration of political contestation in Brazil from their work around Belo Monte? These are the questions to which the book now turns.

6 · THE LAW, ACTIVISM, AND LEGITIMACY

At 9 A.M. on a sunny morning in April 2012, less than a year after Belo Monte's construction had begun, I arrived by motorcycle taxi to a small church on Fish Market Road in Altamira. On one side of the street was the river and on the other was a neighborhood of small wooden stilt homes. Just in sight above the humble structures was a hillside spotted with larger homes, each surrounded by walls, gates, and, undoubtedly, security cameras. In contrast to the quiet Sunday morning atmosphere of most of the city, dozens of people gathered outside the church, and dozens more were congregating in the open-air meeting space behind the church. Neither I nor the eighty other people in attendance had come for a religious service. Instead, we were there for a discussion with officials from the Defensoria Pública do Estado do Pará (Public Defenders' Office of the State of Pará—DPE). This gathering was one of the sixteen meetings that the DPE was holding in churches, schools, and other community gathering spaces in the urban neighborhoods of Altamira that would be directly affected by dam construction. Like all of the communities that the public defenders would visit, most of the residents in this particular neighborhood would be displaced and relocated to urban resettlement communities within two years. The public defenders hoped they could clarify the compensation options available to residents and business owners and convince residents that the DPE could be a crucial ally for the community before and during the resettlement process.

I made my way to the meeting space behind the church, which was covered by a tin roof and open on three sides. The back of the church served as a wall on the fourth side. About ten wooden chairs were placed in front of fifteen short wooden benches, leaving a few rows worth of standing space behind the seating. On a small table in the front of the area sat a computer and a water bottle. Hanging on the front wall was a banner with the logo of the public defender, which included the letters "D" and "P" in bright red block script, a small blue star overlaid on the middle of the DP, and the full title of the agency, Defensoria Pública do Estado do Pará. I greeted a university student who worked for the public defender and who had invited me to the meeting, and he introduced me to Andreia Barreto.

Andreia looked to be in her mid-thirties. She was wearing glasses, a watch, a light pink dress, and white shoes with heels. This professional outfit made her stand out from the residents of the neighborhood, most of whom wore T-shirts, shorts, and sandals. She introduced herself with a warm smile and spent a couple of minutes with me, asking me where I was from and what I was studying before turning her attention to preparing for her presentation. She spoke quickly and matter-of-factly but was pleasant and welcoming. Over the subsequent years, I would come to know Andreia well. She was the public defender based in Altamira who devoted most of her time to issues related to the dam and its effects on residents. She attended and organized a wide range of meetings with and about dam-affected communities, and she assisted people throughout their claims-making efforts, such as the fishers' protest and occupation.

In the fifteen minutes before the meeting began, the area filled with men, women, and children. I stood next to Antonia Melo da Silva, the leader of Xingu Vivo, toward the back of the space, against one of the wooden poles that supported the roof. This part of the church was built on stilts, and water slowly flowed underneath us. To my right were three houses whose wastewater ran directly into the river below, which was also filled with floating trash. Two women were standing on planks just outside their homes, hanging laundry. As more people filtered into the meeting space, a few kids peaked out of the window of one of the wooden homes to see what was happening at the church. Behind us was an open, grassy swamp area about a hundred yards across to more stilt homes and a large dirt area, on which a group of young people were playing soccer. Just a few years later—after the completion of the resettlement program and city development programs that accompanied Belo Monte's construction—the church and the stilt homes would be gone, and much of the remaining land would be a park.

The meeting began with Andreia and Artur, another public defender, discussing why they were there. "First," Artur began, "we are here to uphold our institutional mission, which is to guarantee free legal assistance to those who need it." He continued by explaining that the DPE had created a special working group to support people affected by the dam's construction in new ways. Instead of their typical method of finding clients, in which they wait for people to approach them with needs, the public defender was approaching the affected neighborhoods. Artur explained that they wanted to ensure that the people directly affected by the dam would know what services were available. Andreia continued by saying, "We want to offer a proposal to work together. We are here with some information to share, but we want to hear your questions, opinions, thoughts, and experiences so that we can work together in the future." She stressed that community members had the right to participate in the process. The public defenders, she explained, would do everything in their power to ensure that right was

FIGURE 6.1. A meeting the public defenders' office hosted in April 2012 with urban residents who would be displaced and relocated. (Photo by author.)

fulfilled, so that people could receive compensation for the loss of their homes or businesses.

After speaking for about thirty minutes, Andreia concluded the presentation by asking for an informal commitment from the community. "Will you accept our proposal for the public defender to work with you on housing-related issues?" A resounding "Yes!" filled the air in response. Andreia and Artur then opened the meeting for discussion. For well over an hour, the two public defenders responded to questions, listened to residents share their experiences, and discussed how their office could be an advocate for the community in the coming years. One middle-aged man stood, and with a tone of frustration but not quite defeat, suggested that it was not possible for him, his family, or most people in the community to influence how they would be treated during dam construction and resettlement: "The poor do not have a voice in this process!" A short time later, a young woman said she believed the construction consortium Norte Energia and the government would rebuild the neighborhood and sell the properties after the neighborhood of stilt homes was gone. "They will earn big money here. The land here is very valuable." Other people shared similar sentiments. Not only were they sharing the importance of that specific place, they also felt powerless and without enough knowledge of the process to be informed participants. Andreia reiterated that she and her colleagues were speaking with affected neighborhoods precisely to show that there were support structures for them and that citizens had the right to participate in the processes that impact their lives. The public defenders' office, Andreia stressed, could help people navigate the situation in the best ways possible.

The series of meetings with dam-affected neighborhoods in the city was just one of many places the public defender and public prosecutors appeared. Whether I was at a protest organized by a social movement, a public meeting with a dam-affected group, or an event with Norte Energia and government officials, at least one representative from these legal institutions was usually present. This was particularly striking, because the public defenders' office had been open for less than a year, and the public prosecutor was understaffed. What role were they playing? What was the relationship between the public defender, public prosecutor, social movements, and dam-affected populations? How did they manage their relations with activists, government officials, and Norte Energia? Could they actually make a difference for people most impacted by the dam's construction, and if so, how could they do so?

These types of questions have been central to research on when, why, and how the law is mobilized by society—and the effects of such mobilization—in contexts of injustice and social struggle globally.[1] Much of the scholarship on the use of law and social movements has focused on the role of litigation and the use of the courts. For decades, many scholars have argued that mainstream legal institutions reinforce inequalities, support elites, and prevent social change, rather than support movement goals.[2] This critical assessment continues, but scholarship has also begun to show myriad ways law impacts movements and the broader political terrain.[3] The forms of engagement that public prosecutors and public defenders embraced before and during Belo Monte's construction show the importance of moving beyond a focus on litigation. This case deepens our understandings of how lawyers and legal institutions influence social struggles.

Throughout Brazil, legal institutions and the rule of law have emerged as central mechanisms of claims making in a range of social and environmental conflicts.[4] The case of Belo Monte shows how and why people use the law, as well as the impacts of lawyering on legal institutions themselves and the broader political realm. For example, the public defenders' use of original outreach strategies, such as the neighborhood gatherings, pushed the boundaries of their institution. The public defenders refused to wait for cases to come to them, as the preexisting limits of their work dictated. They were approaching affected neighborhoods to ensure that the people directly impacted by the dam would know what services were available and to help build trust with the communities they felt they could support. In addition, they were creating alliances with social movements, activists, and nongovernmental organizations. By doing this boundary-pushing work, Andreia, Artur, and their colleagues were striving to become principal allies of marginalized communities impacted by the dam. The public defenders' innovative outreach and other strategic interventions were also attempts to strengthen their organization and legal institutions, more broadly.

The ability of officials from these legal institutions to employ their strategies to navigate the social, political, and institutional changes in Altamira were largely due to their ambiguously defined positions. The public prosecutors' and public defenders' offices are state-created institutions that bring claims against the state, support citizens that are harmed by state actions, and partner with activist organizations. They are at once a part of the state and also autonomous from direct government intervention. They have legal roles and responsibilities but are often viewed as activists. This ambiguous nature allowed them to be seen as legitimate mediators of claims in multiple contexts, as they were able to creatively expand their functions and respond to the needs, voices, concerns, and opinions of civil society, the state, private enterprises, and dam-affected citizens. The ambiguity of their position also enabled these legal institutions to function both within and outside of the judicial system. In fact, both the public prosecutors and public defenders often made more progress on cases in which they served as mediators without entering traditional judicial processes and without using the courts. In most cases, these lawyers in Altamira were able to use the ambiguity to their benefit, but they also faced structural challenges that made their work tenuous and potentially unsustainable.

BUILDING LEGITIMACY AND FILLING KNOWLEDGE GAPS

The public defenders' attention to Belo Monte was both important and difficult for the institution. The organization wanted to establish itself as a vital public service provider that offered high-quality legal services and support for people facing challenges due to infrastructure projects, such as dams. The DPE had created a special working group to address issues related to Belo Monte, but some people questioned the institution's ability to be a credible and relevant actor in the conflict over the dam. For example, in May 2013, a public prosecutor in Belém regretfully explained in an interview that the public defenders were not in a position to support the people suffering due to dam construction. He explained that, despite their hard and important work, it was a weak institution with too many problems, too few people, and too few resources. He pointed to the lack of infrastructure within the court system in Altamira and said the state of Pará lacked the resources to resolve the problems. His comments came well after the state had begun increasing funding to the public defenders' office in the mid-2000s. Despite the subsequent improvements in the organization, the legacy of the DPE as an ineffective institution that lacked resources plagued their reputation—even among people who supported them—and made their work more difficult.

Andreia and other public defenders recognized the institutional challenges that could impact their credibility and legitimacy. They sought to utilize the moment to establish themselves as a relevant and effective defender of citizen rights. Very few public defenders throughout the country had experience working on cases in which significant portions of the population are affected by a large infrastructure project. The public defenders in Altamira believed that the work they carried out in that context would impact the future of the institution in Pará and perhaps throughout the country. During the first two years of construction, Andreia told me on a few occasions that what they managed to accomplish—and the way they went about doing it—would influence the work of public defenders in future projects. During one meeting, she explained, "How we act and deal with this case will have implications in other situations, such as the Tapajós [dam complex], where there is a public defenders' office, but where they are in the early stages of discussing human rights." This position motivated experimental and imaginative work that pushed the boundaries of the institution, such as the urban neighborhood meetings.

This inventive work was challenging and complex, as communities lacked clear information but were hearing differing things from government officials, representatives of Norte Energia, and researchers. These dynamics further complicated the public defenders' outreach efforts in riverine communities. A few months after the meetings with urban neighborhoods and just prior to the fishers' protest and occupation of the dam site, the public defender visited the rural riverine communities that would be impacted by dam construction. The first boat trip was to a community on the banks of the Xingu downstream from Altamira and upstream of the dam. Andreia had been excited to talk with the riverine people, as she knew they faced impending changes and likely knew even less than urban residents about the process. She assumed this trip would be of interest to me, so she invited me, a university student who worked in the public defenders' office, Antonia Melo, and two residents of the riverine communities. We met at the main port in Altamira early on a Wednesday morning and were soon heading away from Altamira and down the Xingu.

After an hour, the boat passed close to shore and then turned around to enter a small natural canal, created by a few large boulders. On the other side of the rocks, the water was calm, and we pulled up to a small sandy shoreline. Only one small wooden house and a cattle fence were visible as we disembarked. Nobody seemed to be around, so one of the former community residents who had traveled with us led us up a small hill. We crested the hill after walking a few hundred yards along a dirt road and soon reached a very modest two-room schoolhouse made of simple wooden slats and a tin roof. The yellow-and-green paint on the outside walls was badly chipped and wearing thin. Inside, about ten children under the age of ten looked toward one end of a large room. They were paying attention to a man who appeared to be in his seventies. The man was standing between a simple wooden desk and an old chalkboard covered in old tape marks and water stains. A single bookshelf, which held a few dozen textbooks, stood in the corner of the room, and about fifteen desks were scattered in the room. The alphabet was displayed above the chalkboard, with each letter on a white piece of paper that looked to have come from a student workbook. Felt letters were glued to the wall to spell "Sejam Bem Vindos," Portuguese for "Welcome." Light shone through the small gaps between the boards of the exterior walls, and two large wooden windows were open to provide airflow. The yellow walls and green windows gave some life to the otherwise barren room.

Our arrival distracted the kids, who all looked in our direction, and the man stepped out of the building to meet us under the small overhang that protected us from the sun. The man's white, collared shirt with a green vertical stripe down the left side and the logo for SEME, which stands for Municipal Secretary of Education, indicated that he was the teacher. He wore athletic shorts and sneakers, which seemed to belie his age. The teacher told us that the residents knew about the meeting and would be arriving in about an hour. Andreia wanted to take the opportunity to walk around the community, meet the residents, and hear their stories.

We decided to walk back down the hill to the house next to the shore. Outside were large fruit trees and beds of vegetables that were raised five feet off the ground. Inside, Andreia's assistant set up a tripod for a video camera. Andreia and Antonia proceeded to ask the owner questions about her household and the community. They wanted to know what the family grew on the farm, how many kids lived there, the challenges in the community, and other aspects of their livelihoods. The woman explained that they sold a variety of other crops in the neighborhood and in the city. All the kids in the community went to school here through eighth grade, and some then traveled to the city to continue their education. Education in the city was difficult at the time, however, because most of the schools were overcrowded due to the influx of people coming to the region for work on the dam.

Andreia then asked if Norte Energia had registered the household and property. Prior to the resettlement process, representatives from Norte Energia or a subcontracted company should have visited the community to document who lives in the households and evaluate the property. Norte Energia, along with input from government officials, had created a "notebook of prices" that indicated how much a family should be compensated for each item. For example, each cocoa tree was given a price based on the age of the tree. The inspectors who visited households during the registration process should have documented all of the crops, buildings, and other items that would be lost due to displacement. Families would then be offered a compensation package, which might include cash or a specific place to live. Officially, communities could request to be resettled as a group to a location that would have similar characteristics as where they were moving from, but many individuals did not know or understand this process. Families would instead sign individual agreements. In this case, the woman told Andreia that some sort of registration happened but she was not sure what she had signed. Andreia asked her name so that she could document the details and presumably follow up once back in the city. The woman provided a first and last name. When Andreia asked if her given name was longer, which would be common in the area, the woman was not exactly sure, so she retrieved a backpack that was packed full with paperwork. Together, Andreia, Antonia, and the woman looked through the backpack to find her official documentation. They concluded the conversation after sorting through who she was and finding the documents that she had signed for Norte Energia. The woman clearly did not understand what she had signed or when she would be moved. This would be a topic of conversation with the community at the meeting.

We made our way from the house back to the school. As we arrived, I saw over ten men talking with each other outside and an equal number of women inside. Given the isolation of the area, some of the people had traveled from a few kilometers inland to attend. As Andreia and Antonia set up the room for the meeting, I struck up a conversation with the teacher and some residents. The teacher, who had recently come to the area to begin teaching at age seventy-one, said that the community was extremely tight-knit: "Everybody knows everybody really well. We had a party here last Saturday and everybody was here. Fights are rare." People were displeased at the thought of moving to places far from the neighbors they knew best. They were also angry and confused as to why they were not receiving benefits like some of the Indigenous groups who lived in demarcated "Indigenous" land, who were guaranteed financial and other material compensation through the mandated mitigation programs. The teacher explained that most people who live in this community are partially Indigenous, but do not receive any benefits for it. One woman chimed in to say that her mother was 100 percent Indigenous but her father came from the city of Altamira.

This type of story was not uncommon. Indigenous people with mixed ancestry and those who did not live on demarcated lands were in ambiguous positions in regards to compensation. Many lived along the river and relied on fishing to survive, yet were not officially recognized as part of a dam-affected group. They may have been Indigenous but were not given the same financial benefits or legal protections granted to Indigenous communities. These dynamics created resentment, anger, frustration, and confusion, and influenced how people made demands. Some residents in this situation claimed their Indigeneity during claims-making processes, either explicitly in their statements or implicitly painting their bodies in Indigenous symbols or by wearing traditional Indigenous attire, such as headdresses or beaded necklaces. While nobody wore such apparel while we visited this riverside community, the politics of indigeneity was nevertheless apparent.[5]

The meeting began shortly thereafter, and the room was lined with a couple dozen people listening to the teacher introduce Antonia and Andreia. Antonia began by giving a history of the movement to stop dams from being built along the Xingu. She transitioned to talking about the current moment, saying that she understood that people there knew little about what the impacts the dam would have on their lives and the process through which they should be compensated. She stressed that the movement was there to support rights of riverine communities and that a variety of people from Brazil and all over the world were there to help, pointing to me as an example. She mentioned the federal prosecutor as another important ally before introducing Andreia.

Andreia provided a similar introduction and explanation as she had given to the urban neighborhood communities, explaining the judicial process and how the work of the public defender can be particularly helpful during the displacement process. She wanted to orient the affected families to what was happening and what should happen, to question whether Norte Energia had been doing their registration work in fair and ethical ways, and to build the community's trust. As in the meetings in the city, she stressed that she wanted and needed to learn from them in order to support them in the best possible ways. With that goal in mind, she devoted much of the meeting to listening to the experiences of the community. "Who here has been registered?" Most of the hands in the room went up, but she didn't know exactly what that meant to each of them. "Have you signed anything?" A few people said that yes, they had signed some papers, while others had been told that people would be back to continue negotiations. Andreia then said it would be helpful to see what they had signed, but nobody in the room had been given a copy of any signed documents. She explained to them that this was an example of a serious error committed by Norte Energia: "You have a right to have a copy of what you signed."

This sparked a long conversation about their rights and responsibilities before and after agreements were signed. They discussed problems with inconsistent and undervalued property evaluations, the timeline moving forward, the challenges in actually receiving money for the land, and the options that the families faced in resettlement. Andreia also strongly suggested that people continue to work the land on which they lived; otherwise, the property value may decrease upon another inspection. People were confused by this because it was difficult to invest in land that they were fairly certain they would be forced to abandon. The group continued to discuss these types of challenges for well over an hour. Andreia and Antonia reiterated the importance of solidarity and the support that the public defender and social movements could offer communities in need. Antonia invited everyone to a rally that would take place in the city the following weekend, on Independence Day. She was trying to recruit people to the anti-dam movement. Andreia offered to return to hear and document all the individual cases in the area.

It became clear during the meeting that many of the residents were unaware of their rights and what would happen in the coming months and years. It was also clear that the residents were reluctant to start working with the public defender immediately. Andreia and Antonia were well aware of this challenge. After the meeting, the two of them explained these challenges in supporting dam-affected

FIGURE 6.2. Family washing clothes on banks of riverine community that would be impacted by dam construction. Boats pictured in photograph are typical of those used by fishers in the region. (Photo taken by author during a visit to the community with the public defenders' office in September 2012.)

communities. Antonia remarked, "People here do not trust anyone anymore." Andreia added, "We have to deal with these issues in a delicate manner. It is a slow process and people do not have much information." The challenge in building trust came from the relative isolation of the community prior to Belo Monte's construction and the sudden deluge of people who began passing through. Representatives from Norte Energia, subcontracted companies, and research firms had begun visiting the residents' properties, asking questions, assessing the land, showing documentation, and asking residents to sign documents. Few of these residents had the means to assess the validity of the claims anybody was making. They understood that this process would provide some compensation for the land they would soon lose, but had few details of their rights and what they were owed. Some people in this situation had signed agreements to receive monetary payments for their property, and so were wary of outsiders interrupting the process of compensation. Andreia and Antonia were well aware of the confusion, so were careful in the ways they addressed people.

The meetings in both the urban neighborhoods and riverine communities show the types of creative outreach the public defenders engaged in and the reasons why they did so. In the rural areas, people were isolated and often without adequate support structures. In urban areas, the city had changed quickly, and there was a great deal of misinformation. In both areas, the arrival of new institutions, subcontracted companies working on the dam project, activists from outside the region and even outside Brazil, and researchers led to an overabundance of sometimes conflicting information. People lacked reliable knowledge about the impending changes that would drastically affect their lives. As people struggled to know what to do next, legal organizations, and in these cases the public defenders' office, were trying to step in as vital institutions for the future.

ACTIVISM AND THE LAW: BUILDING NETWORKS, BLURRING BOUNDARIES

About a week after the meeting at the urban church, the public defenders' office hosted a gathering in another neighborhood whose residents would be displaced. The meeting took place in a two-story green-and-white school with a large auditorium, which was used for sports and events. As I arrived, I once again saw Antonia, the Xingu Vivo leader. As we walked up the stairs to a large classroom on the second floor, she explained the origins of these outreach meetings. Antonia suggested that she developed the idea the previous year with other activists in Xingu Vivo and then credited Andreia for making the idea come to fruition. It was becoming clear that an important network was forming, and the alliances between social movements and legal institutions would impact both sides, as well as the people they were seeking to support.[6]

The importance of the network was on display in meetings that the public defender held with the communities. As Antonia and I made our way down the hall of the school, a small group of young people were moving chairs into the large classroom. Antonia and I helped, and the seating filled almost immediately. When the meeting started ten minutes later, about 125 people were in the room. In an effort to leave space for the neighborhood's residents, I stood against the back wall. Andreia and Artur gave a similar introduction, but this time they projected a series of slides on the front wall to articulate their points more clearly. As Andreia finished introducing the work of the public defender, she stressed once again to the audience that residents—particularly residents directly impacted by dam construction—had the right to participate in the process, to know what was happening, and to make claims for better, fairer compensation if they felt Norte Energia was not being forthright. During the question-and-answer period that followed, Antonia took the opportunity to support Andreia's comments, imploring people to participate. She did so by speaking in a direct manner, adding comments that would have been inappropriate for the public defender to say. "The government, Norte Energia, and other companies always lie to the people!" Antonia exclaimed. "They do not consider the rights of the people affected ... We need to study what is happening. We need to organize. We need to participate!" A leader of another movement, in the audience, made similar points, stressing that people needed to get involved, use the public defenders' services, and participate in the social movements that were fighting for people's rights: "Social movements have a strong history in this region of forcing the government to take care of the people. Now, the public defender is doing something that is different than, but compliments, what the social movements are doing." This activist, along with most others in Altamira, seemed to recognize that the traditional strategies of collective action were not as effective as they once had been on their own. The public defender was fast becoming an important addition to the network of actors working to support the most marginalized and was using other institutions to navigate the complex changes of the moment.

The active support from social movement leaders offered a different kind of voice for public lawyers and made evident the alliances and mutual support between the public defender and social movements. These activists were sending the same messages as the public defender but could make comments that were inappropriate for the public defenders to say, given their institutional positions. The public defenders were a bit more reserved and based their comments strictly on the facts. They focused on the role that they could play in supporting the population, while the activists used stronger language about the "crime of Belo Monte" and offered other ways for people to get involved through rallies and community groups.

These dynamics also played out during the visits that the public defender and activists made to riverine communities. On one trip, they traveled to rural areas downstream from the dam, places that would confront unique and complex challenges. The residents there would not be displaced, as their land would not flood, but they would face significant changes nonetheless. River levels would decrease for most of the year as water was diverted upstream for the dam, and navigation by boat would be disrupted. In addition, a Canadian mining company, Belo Sun, had introduced plans for a large gold mine in that section of the Big Bend. These plans would become the source of a heated debate in subsequent years, but people knew little about it at the time. The public defender and social movements were using these meetings to collect information and encourage these communities to stay engaged. All of these outside groups felt they could support the residents.

At one meeting, the public defender explained the known details of the mining project and the rights that people would have before and during construction. Antonia then stood up and, without mincing words, argued that the mining project would have grave consequences: "With Belo Sun, everything as you know it here will end and everyone will be affected! But all of you have rights!" She shared stories from communities removed during the early stages of Belo Monte's construction and then exclaimed, "You cannot accept what they offer you. The force of the people is powerful. You are powerful!" The two groups were again using each other to share similar points, but doing so in different styles.

In contrast to the still unknown public defender, residents of the region recognized the activists and social movement leaders. After decades of fighting for rights, the activists were known in the community as people who had helped make gains for the region. The activism struck residents in different ways. Some people were encouraged by what the movements had accomplished. Others were turned off by the militancy of some of their tactics and their dogged resistance to Belo Monte and Belo Sun, which some people felt were development projects that could bring much-needed attention and resources to the region. Regardless of the perception, residents recognized these activists as supporters of the community, and their presence helped shape perspectives of what the public defender was attempting to accomplish.

As these outreach efforts continued and as Andreia participated in Xingu Vivo's meetings and rallies, I began seeing how the public defender—and public legal institutions more broadly—were taking on activist roles, even as they attempted to maintain legitimacy as independent legal organizations. Some scholarship on cause lawyering, or the use of law to promote social change or support a particular social cause, has critiqued this type of work for how it can create elite politics, leave out many of the people whom the lawyers are attempting to defend, challenge the professional moral neutrality that many lawyers pledge, and simply be rendered ineffective.[7] In contrast to those critiques, the public defenders were attempting to do the opposite by consciously making their work less elitist and more grassroots, embracing activism and blending it with their professional identities as lawyers.[8] As traditional forms of engagement weakened, the officials in legal agencies emerged as voices of authority with robust institutions behind them. Social movement leaders and officials from the public legal institutions were blurring—but still maintaining—the boundaries between activism and legal action. They had built alliances yet attempted to serve the communities in different ways. Antonia and social movements wanted the support and the information that the public defender and others could offer, but by no means did they want to be beholden to the slow, bureaucratic functions of the legal forms of claims making. Andreia, on the other hand, was committed to maintaining integrity, so as to not jeopardize her reputation with judges, representatives from private entities, and government officials.

Andreia and I occasionally talked about the dynamics between activism and legal forms of recourse. Toward the end of my intensive fieldwork period in 2013, I asked Andreia if she viewed her work as activism. She looked up from behind her desk, smiled a bit, and admitted that yes, it was a kind of activism. Andreia argued that everyone involved created and used certain images and identities to make demands. For example, many Indigenous groups in the Xingu region received significant financial and other material benefits during dam construction, as compensation for the ways the dam impacted their lives. In this context, Andreia argued, if somebody claims an Indigenous identity in hopes of receiving something in return, they are using the image of indigeneity as a tool for claims making. Others claim to be activists and gather the support from social movements.

She explained that she uses the institution of the law to make demands: "Some have guns, some have an image, and others have institutions. As an activist, you use what you have."

MEDIATING, NOT LITIGATING

The work of the public defender in the urban neighborhoods and the rural communities helped pave the way for slow but significant advances that the fishers and riverine people made in the subsequent five years, even though the public defenders' office of the state of Pará temporarily closed in 2015 due to a lack of funding and as a result of decisions at the state level. Thais Santi, the public prosecutor who focused on issues related to Belo Monte, took up the case of the fishers, continuing the innovative work that Andreia and the public defenders' office had begun. Thais's work with riverine communities is peculiar both for its relative success and its unconventional path. She made more progress for riverine communities outside the courtroom than in it. Neither she nor Andreia brought an official legal case on behalf of the riverine people, who faced many changes to their lives. Instead, these officials built networks of support and used those networks to create more information and knowledge about the plight of river-based communities. In contrast, the cases they litigated made no progress, with Norte Energia digging in its heels and using ample resources to mount strong legal defenses.

The progress that Thais made was far from immediate, and the trust that residents put in her took time to develop. Thais arrived in Altamira in the middle of 2012, a few months before the fishers' protest. At the time, Andreia and the public defenders' office were carrying out the meetings with riverine communities and had built enough trust to represent the fishers during the long protest and occupation of the dam. Thais, in contrast, was from a state in southern Brazil, had no connection to Altamira, appeared elite, and was just beginning to learn about how people lived in the area and the impacts of the dam. Thais had voiced support for impacted communities, such as the fishers and Indigenous people, but some people were skeptical. Her background and unfamiliarity with the region made some activists and others wary about Thais's ability to support the most marginalized.

These suspicions were heightened during the conciliatory hearing that brought the fishers' protest to a close (detailed in the prologue). During that two-day meeting on the construction site, Andreia and the public defenders' office supported the fishers, while Thais served as a facilitator between those occupying the dam site, government officials, and representatives of Norte Energia. Throughout the meeting, Thais was careful not to play favorites, even commenting that she was not representing any of the groups at the meeting, including the Indigenous people or fishers. It was a tense environment. Construction had been halted in the area for over a week, and Norte Energia was anxious for work to resume. The people who had been protesting for weeks were worried they would not see any tangible results. Thais's approach to managing the tensions of this situation was calculated, as she attempted to help all sides come to a resolution. She made sure everyone had opportunities to speak and focused on responding to complaints about the process. While she had previously voiced concern for the fishers to me, she did not publicly advocate for the rights of any of the groups who had been protesting, which did not sit well with some of the regional civil society and social movement leaders.

The questions of Thais's allegiances would be fully allayed over time, as the stories in the previous chapter show. She would become one of the staunchest voices for the people who fished for a living, those who lived on the banks of the Xingu, and other marginalized communities impacted by the dam. She built relationships with a range of groups, including the public defenders' office, nongovernmental organizations, and social movements. She developed teams of researchers in efforts to give expert support to the causes she took up, and she was not afraid to file cases against Norte Energia, the government, or private actors. When the public defenders' office of Pará closed, the resettlement process was about to get underway in the city. Thais recognized the lack of support, legal and otherwise, that existed for the residents who would be displaced, so she used her institutional connections in the country capital, Brasília, to help open a federal public defenders' office, the Defensoria Pública da União (DPU).

The process of building institutional legitimacy was easier for Thais than it was for Andreia and the public defenders' office. The public prosecutors' office was well known throughout the country, and many knew that some prosecutors blended their professional work with activist inclinations. Additionally, Andreia had done much of the work in the region to build legitimacy for the use of creative outreach strategies by legal institutions, from which Thais benefitted. Furthermore, while Thais's facilitation at the conciliatory hearing may have temporarily slowed the building of trust with some dam-affected communities, her balanced approach also helped strengthen her legitimacy in the eyes of Norte Energia and government officials. Officials from the government and nongovernmental organizations, as well as the companies that oversaw Belo Monte, soon saw Thais as a personal advocate of marginalized communities with legitimate authority to support those communities.

The people impacted by the dam also came to see her in a similar way. In fact, fishers and riverine people credited Thais with a great deal of their progress. For example, nearly everybody at the celebration of the one-year anniversary of the Riverine Council's formation publicly stated that their success would not be possible without Thais. In private, people also lauded her and other legal institutions. Rodrigo, the informal leader of the fishers' protest and founding member of the Riverine People's Council, who had been relocated from his home on the river to a house five kilometers away, explained how the majority of the council members needed people and organizations to guide them: "The Riverine Council is made up only of fishers and riverine people, people who haven't studied, and the majority of us are not literate... We need someone who can guide us, lead us." He discussed the invaluable role of nongovernmental organizations, legal institutions, and social movements. "The public prosecutors' office, in Doctor Thais, was the one that gave us real potential," he said. "The public defender, Xingu Vivo... these are the entities that are really guiding us."

Jose, the member of the Riverine Council who had become emotional during the celebration at Gloria's property, shared similar thoughts. During an interview at his house, Jose gave a great deal of credit to Thais for the start of the council: "She knows our reality. She has learned our reality." He complained that the people responsible for Belo Monte only pretended to know his way of life. "How do people from Brasília and São Paulo know how we live?" he asked. "They have studied this area to be dammed for forty years, but not the right way. It's only us who know what it is to be a riverine person." After he seethed about the lack of understanding and respect for their way of life, he returned to talking about Thais: "So, how are our rights defended? Thais looked for a better way. She is a person who will fight for our rights. Others won't but she fights for us." Thais clearly meant a lot for the people, particularly the fishers and riverine communities, who struggled for their rights in the face of Belo Monte.

The legitimacy that Thais and Andreia built among residents, government organizations, and private companies turned out to be crucial because of the surprising avenues by which legal institutions were able to make gains for dam-affected communities. The only progress that Thais, Andreia, and other lawyers made on the case related to riverine people—or any case, for that matter—was accomplished outside of the courts and official legal channels. These legal officials arranged meetings for the riverine people to speak about their experiences, organized research to support the cause, and negotiated directly with the local government and Norte Energia for compensation. This progress was in stark contrast to the official legal cases, which were often held up in court. In fact, one of Rodrigo's motivating factors for starting the Riverine Council was the slow progress of a legal case. He explained that the public defender had filed a case related to problems with the registration process of displaced people. "For two years," he said, "I lost time going to check on the case. [I would hear] 'The case is not progressing.' And then 'It is progressing.' And nobody protested! After two years we started to say... 'If we do not organize to fight for our rights, we will stay in this situation.'"

The lack of progress on legal cases was a common story. In 2017, I interviewed Tomas, one of the federal public defenders, in his Altamira office. On a table next to his desk were stacks of pink folders, each representing an official claim related to Belo Monte that their office had brought before a federal judge. Tomas explained that all of the cases had been halted because the judge deemed them to be under the jurisdiction of the state courts. Most of the cases were at least two years old, he explained, and revolved around complaints over the urban resettlement process. Housing was one of the most precarious and challenging issues that people faced in the city, and many people felt they had not been compensated fairly or were facing unjust conditions in their new neighborhoods and homes. Tomas was planning to address the concerns and hoped the judge would change the decision, but he was far from certain that the cases would even be heard.

The challenges of traditional legal claims and the fact that mediated negotiation often replaced those legal cases highlights another way deliberation was becoming the primary form of making demands and bringing conflicts to a resolution in Brazil. Along with state-controlled participatory mechanisms, such as CGDEX, and the new opportunities for participation initiated by the fishers, legal institutions had begun to be promising means of direct negotiation. Public prosecutors and public defenders established themselves as mediators for the claims made by populations impacted by the dam. Rather than rely on the rule of law to adjudicate claims, creatively minded legal officials used all of their resources, networks, and institutional authority to convince Norte Energia officials to remedy injustices they had inflicted on marginalized people.[9]

While this shows the opportunities that are at the disposal of legal institutions, this case also highlights structural barriers in the legal system and points to the vulnerability of people most affected by the dam. Through the many cases related to Belo Monte, the legal system was shown to be a clumsy and slow process that could not offer support in a timely manner to the people who most needed it. In addition, the success of the unofficial negotiating strategies in this case leaves open the question of whether or not the institutions, rather than the individuals like Andreia and Thais, could have effectively supported the population. The officials in these institutions took a great deal of initiative in supporting people through inventive mechanisms. Had other public defenders held these positions, there may not have been as much progress. In addition, the legitimacy they built over time was incredibly important, yet also tenuous. Furthermore, these officials sometimes had their hands tied by their agencies. Andreia had to leave Altamira when the public defenders' office closed, just as thousands of people were being displaced and resettled. The federal public defenders' office picked up some of the cases, but was slow to open an office and had other structural problems. For example, all of the federal public defenders in Altamira were on their first assignment after their studies and were reassigned quickly. Altamira was not a desirable location for most young public defenders, who preferred to be in a less isolated place. The quick turnover hurt the ability of these officials to build trust and rapport with the populations they were serving and with other agencies. Together, these challenges suggest that these innovative, nonformal strategies may be difficult to replicate, are lacking some structural support, and are tenuous approaches to claims making. Nevertheless, legal institutions became central actors in political conflict in the region.

AMBIGUITY AND THE CHALLENGE OF BEING SEEN AS AN ACTIVIST

The public defender and public prosecutor pushed institutional boundaries and developed trust with the populations they hoped to serve, but their work also had challenges. On the one hand, their reputation as activist organizations increased their legitimacy among members of the most outspoken social movements, and their work provided some limited, varied support to dam-affected population. On the other hand, these legal agencies had to manage and consider relations with government and private-sector actors. Some government officials and private-sector representatives—who were often the targets of the demands—did not always view the activist identity in a positive light. The challenge of managing these relations was both heightened and mitigated by the ambiguous nature of the legal institutions as public, state-created organizations that were tasked with holding the state and other powerful actors accountable.

Norte Energia representatives, in particular, called the legitimacy of the legal agencies into question whenever they had the opportunity to do so. Their justifications for objecting to aspects of the work of legal institutions were clarified when I met with Rafael, Eduardo, and Elena in 2012. Rafael was a high-ranking employee of Norte Energia, and I was expecting to interview him alone when I arrived at the headquarters of Norte Energia, which were on the outskirts of the city. At the time, it was notoriously difficult to arrange meetings with managers and directors of Norte Energia; they distrusted outsiders. These officials often conveyed information to the public through communications teams, which could carefully construct messages and images. I met members of these teams throughout my time in Altamira, as well as other people who served as liaisons. Eduardo, for example, worked as "interface coordinator" for the Consórcio Construtor Belo Monte (CCBM—Belo Monte Construction Consortium). CCBM was the entity responsible for the actual construction of Belo Monte and had a contract with Norte Energia, which paid CCBM to hire and manage the contractors to carry out the work. In his role, Eduardo acted as one of the liaisons with the public, arranging site visits to the dam and managing relations between CCBM, the public, and political actors in the region. He had arranged the meeting for me, and when I arrived, he explained that he and Elena would be joining me. Elena was an employee of an independent company that oversaw communications for Norte Energia and CCBM. It seemed as though Elena and Rafael wanted to manage the conversation and control some of what was said to me. As the three of us sat in Rafael's office, we spoke about the projects Norte Energia was carrying out to mitigate the effects of the dam, the relationships between different actors, and the challenges they faced in doing their jobs.

Toward the end of the conversation, I asked Rafael what he thought of the public prosecutor and public defender. "It's part of the process," he explained. "The public prosecutor has their function. The public defender also has their function. I just think that what cannot happen in the process is when the technical turns political... when the technical reason for an argument becomes [an individual's] political reason—" Here Elena interrupted him in an attempt to clarify: "I saw various meetings and hearings where the public defender spoke out against Belo Monte. It's exactly in this way that she comes from a political position." Eduardo added his own thoughts: "A person can have this political position but not in her function with the public agency. That has to be technical."

In their responses, Rafael, Elena, and Eduardo attempted to distinguish the "technical" from the "political" and questioned the legitimacy of the officials from the public legal institutions. This attempt to delineate the political from the technical was a fraught endeavor. They were employing "political" to signal the use of perspectives that can be disputed or that incite conflict.[10] The technical, for them, indicated impartiality, based on established and agreed-upon facts. The distinction is problematic because what constitutes a technical matter and the details of technical matters are usually contested, as evidenced by the debate over if and how the dam affected fishers and riverine communities. The Norte Energia–affiliated employees' social positions—as representatives of a public-private consortium that was regularly in conflict with the public legal institutions—influenced what they saw as technical and whose knowledge they accepted. In attempting to distinguish political and technical behavior, they exemplified how, as David Mosse argues, expressions of knowledge can be manifestations of social relations, rather than expressions of true expertise.[11]

It was no surprise that employees of Norte Energia would find problems with Andreia's approach; after all, the lawsuits filed by the public defender and public prosecutors in this case were often against the construction consortium. Furthermore, these cases threatened to delay construction and force Norte Energia to spend money and use its resources. Nevertheless, the employees' arguments against the activist nature of Andreia's work raised important questions regarding the role of the public defender, the limits of legal institutions in cases such as Belo Monte, and the legitimacy that public legal institutions gain or lose through their inventive approaches. Andreia, Thais, and other public lawyers did not mind that company officials criticized their approaches, but they also wanted to maintain their legitimacy and thus their ability to negotiate on behalf of people impacted by the dam.

The innovative approaches and the activist identity of the public defenders and public prosecutors had mixed impacts on their relationships with government agencies. On the one hand, new and fruitful partnerships formed, such as the "gang of women," the name that Andreia jokingly gave to the alliance she built with influential women in the Ministry of Fish and the federal Government House during and after the fishers' protest. On the other hand, in contrast to these new networks, the public lawyers' approach also threatened to irritate potential government allies who wanted to remain at a distance from social movements and activists or who were advocating development projects that public lawyers appeared to be opposing. Furthermore, a central part of public legal institutions' job was to bring cases against the state. One example with the municipal government of Altamira highlights these tensions best. The public defenders' office once had a very strong relationship with the city's legal team. They came together to work on specific cases that interested both institutions, until the public defender took on a civil suit against the city regarding the inability of ambulances to navigate the streets. According to Andreia, the municipality then viewed the public defenders' office as an enemy rather than an ally, even though they were simply doing their job. The working partnership fell apart. Any remaining alliances further weakened when a new mayor took office in 2013. Many people, including the public lawyers, saw him as closely aligned with and supportive of Norte Energia, the target of many cases.

Taken together, the responses to the work of public legal institutions show the complexity and ambiguity of their positions. Public defenders and public prosecutors could engage in strategies that some would view as "being too political." This could risk their legitimacy in the eyes of the state apparatus or compromise their ability to negotiate outside of official legal channels. Instead of foregoing the creative strategies, though, the public lawyers found ways to manage a complex set of institutional changes while building the partnerships necessary to carry out their work as mediators of claims. As some ministries and government officials started to see the public defenders and public prosecutors as real mediators of the claims of marginalized groups, others were also forced to regard their work as legitimate, even if they disagreed with it.

LEGITIMACY IN QUESTION AMONG THE PEOPLE

In 2017, the majority of dam construction was completed. Many dam workers had moved on, and the buzz of the city had dulled after the height of construction. Just a few years previously, it was extremely difficult and very expensive to find housing in the city. Now, there were "For sale" signs on hundreds of buildings, and rent prices had tumbled; my former apartment was a third of the price I had paid in 2013. Every hotel had vacancies, and rooms were once again affordable, as not nearly as many people were visiting the region. Not only had the number of jobs on the dam decreased by the tens of thousands, but construction projects in the city had slowed as outside attention to Altamira dwindled. Unemployment had unsurprisingly spiked. Displacement and resettlement were mostly complete, but the urban resettlement communities were experiencing disproportionate levels of violence and infrastructure challenges, such as insufficient water supplies, an unreliable transportation system, and a lack of community spaces. The public prosecutors and public defenders in Altamira, as well as some of the nongovernment organizations such as the Movimento dos Atingidos por Barragens (MAB—Movement of People Affected by Dams) were working closely with these residents to solve the community problems. The public defenders were also supporting individual families in the resettlement communities whose homes already had structural problems, such as a leaky roof or a wall that seemed to be caving in.

In addition to addressing the lasting consequences of dam construction, many people involved in the debates surrounding Belo Monte also turned their attention toward the long-debated mining project that was proposed in and around the Big Bend of the Xingu, below the dam site. Belo Sun Mining hoped to exploit the gold reserves that small-scale, artisanal miners had been slowly extracting for the previous half century. Two days after I returned to Altamira in 2017, the public defenders' offices hosted a public hearing in a small village that would serve as the base of operations for the mining project. As I had done for many trips down the river, I went to the city's main port to board one of the boats that would be traveling to the event. When I arrived, it was clear that this trip might be a bit more dramatic than others I had witnessed. In addition to Andreia and Tomas from the public defenders' offices, local reporters were there, as were Antonia Melo and a few other activists from Xingu Vivo, a journalist who had done a lot of work in the Big Bend and supported social movements, and two researchers from São Paulo. One of the activists explained that representatives of MAB had gone by car to the event, which was accessible by ferry over the river and a dirt road through the forest. This kind of enthusiasm and interest in meetings in the Big Bend was not uncommon. On previous trips to the same area, such as meetings the public defender held with riverine communities and the trip I took with the president of the fishers' union, there were often other foreigners, reporters, and university representatives who tagged along. What stood out this time were the two police officers that Andreia had hired to accompany us. Before we set off, Andreia quickly mentioned that she thought there would be tension and conflict, but she did not have time to explain further before getting on a boat with Antonia, the police officers, Tomas, and the reporters.

The other researchers and I boarded a separate boat for the journey, which seemed to pass faster than it had prior to dam construction. The level of the water had stabilized above the dam, removing the small waterfalls that had made certain points a bit treacherous. Just before reaching the dam, we pulled up to a dock, disembarked, and boarded a passenger van. Meanwhile, two men working at the site put the boat on a trailer and pulled it by a tractor. We traveled to the other side of the dam and soon boarded the boat again and continued on our way. We

FIGURE 6.3. Dead vegetation on a flooded island in the Xingu River after dam construction. (Photo taken by author in November 2017.)

made our way through the waterfalls that were still intact on the downstream side of the dam. I was struck by the vibrant green vegetation below the dam that was in stark contrast to the mostly dead and gray trees and bushes above the dam.

Less than thirty minutes later, the boat came around a bend in the river and the village that would host the meeting was visible approximately three hundred yards in front of us. From that point, we could see and hear fireworks. The banks of the small village, where approximately five hundred people lived, were filled with people chanting and holding banners and signs. I incorrectly assumed that people were rallying in opposition to the mining project. Before long, it became obvious that it was quite the opposite. I also soon learned that we were not welcome. A few people on the shore held a long banner that read "We don't want NGOs speaking for us," and another listed the social movements, NGOs, and even the local Catholic Church followed by "Go Away!!!" As the boat attempted to slide onto the shore so that we could disembark, a man stood in the water and angrily grabbed the front of the boat. "Who are you? Are you NGOs?" he yelled. He pushed the boat away, commanding us to leave. Our boat captain did a small circle in the water before attempting to reach the shore again. We approached again, and in an attempt to be heard over the chanting townspeople, the other researchers yelled loudly to him that we were researchers. We made it a bit closer and I quickly explained that I was from the United States and that we were just hoping to observe the meeting. He reluctantly let us on shore amid about two hundred people who shouted and chanted around us.

I suddenly understood why Andreia had brought police officers. Most of the townspeople were angry that there was resistance to this mining project, and they blamed outsiders, who they knew would attend this public meeting. It was clear that the residents of the village did not want social movements, nongovernment organizations, and others speaking on their behalf. The conflict surprised me at first, as it seemed to me that the social movements and public defenders were there to support the population, but I soon saw that the community's disdain for social movements and outsiders made a great deal of sense. The local community saw the mine as an opportunity for jobs and a windfall of money, which they believed would help them overcome the difficulties of life in this remote village. The area had been largely abandoned by government officials for generations, and they were physically isolated from larger population centers. Health care services were almost nonexistent, and the schools struggled to provide basic education. People saw the mining project as the only opportunity to bring hundreds of jobs and financial resources to the area. In the eyes of the townspeople and the municipal government, the social movements and other organizations that had been vocal opponents to the mining project were barriers to development. The outsiders were working against their interests. In fact, a recent lawsuit filed by the public defender had temporarily stalled the mining project, so the people in favor of the mine were angry and resentful.

Amidst the chanting, we walked off the beach and through the narrow streets of the town center. People peered out of the windows of their homes and stood in the doorways of the small convenience shops. After walking the equivalent of two or three city blocks, we came to an open outdoor area, where the public meeting would take place. A long table with three microphones was positioned near the front of the area. A white wooden fence lined the space, and a few large trees provided shade for most of the people who would attend. MAB members had arrived before us in their car and had hung a dozen handwritten signs along the fence, each listing one of the "12 Reasons to Be against Belo Sun." The rest of the fence was covered in signs and banners that townspeople had hung. "We Will Not Shut Up!!! We Want Development!!!" "We also want a hospital, health, education." "What will be my future?" "We need help. SOS!"

In the fifteen minutes after we arrived, more people gathered, and a loud disagreement ensued just outside the gate to the meeting space. The man who had pushed our boat away from the shore was standing next to a large man in a cowboy hat, who I discovered was the mayor of the municipality, Senador José Porfírio. The mayor had arrived to the village two days previously from the urban center of the municipality, which is located more than a hundred kilometers north, as the crow flies, a journey that would likely have taken a full day, including two boat rides and travel by land. The people with whom I had traveled thought that the mayor was leading the townspeople's protests against the social movements, as the mining project would be lucrative for the municipality. Later in the week, the mayor organized a bus full of residents to take a twenty-four-hour journey to Belém, the state capital. There, he and the group interrupted and shut down an event at the university about the riverine people and the mining project. Regardless of whether he was instigating the anti-movement stance, much of the community was convinced the mine would be good for the community and were happy to have their elected official speak on their behalf. Here in the village, prior to the meeting, he was part of an argument with the president of the artisanal mining union that opposed the project. The details of their disagreement were inaudible, drowned out by two dozen residents clapping and chanting, "Go Away! Go Away!"

The commotion settled down as Tomas, Andreia, the journalist, and a moderator sat at the front table. About two hundred people made their way into the meeting space. The majority seemed to support the mining project, and the most vocal proponents gathered on the left side of the table and behind it. Some who opposed the project positioned themselves near the right side of the table, while quiet onlookers stood toward the back. The moderator introduced the first speaker, Ana, a researcher and journalist who had covered the region for many years. Ana lived in São Paulo, but traveled to Altamira occasionally to support Xingu Vivo. She managed their website and helped publicize news associated with the movement. Ana began by saying that she had been coming to this area for over ten years. She had a pamphlet that she started to distribute to the crowd. It depicted people who had died due to Belo Monte's construction. The people did not want to hear it. "This is not Belo Monte!" they yelled. "Talk about people who are alive!" They drowned out Ana's voice and soon began chanting, "Call the mayor! Call the mayor." They wanted the mayor to speak, in order to hear someone who supported the pro-mining position. "Let him speak!" they yelled. After a few minutes of this, he made his way to the table, sat down, and waited his turn to speak. Ana spoke for the next twenty minutes, and she was repeatedly interrupted by the crowd. They yelled things such as, "We are hungry for this project!" and "Will *you* bring us jobs?"

The intense conflict, anti-outsider sentiment, and raucous debates and arguments made what happened next surprising. As Andreia took the microphone, the crowd settled down for the first time. The residents clapped for her before she started speaking. A man next to me said in a hopeful tone, "Now we will see." The crowd knew that Andreia had called the meeting in the first place, and they also knew her from previous visits to the village. After she said "Good morning," most people enthusiastically responded in turn: "Good morning!" Andreia then commented that she has been there many times in the past, that she felt welcomed, and that she knew many people there. For nearly twenty minutes, Andreia spoke about the work the public defenders' office had done in relation to Belo Monte, the phases of the mining project, the rights that people in this area had during each of those phases, and the role that the public defenders' office could play to support them. She explained that the public defender was here to offer legal support for those who could not pay. As she talked, most people were paying close attention, and even clapped for her a few times. Like many of the other meetings she held with communities affected by these large-scale projects, she stressed that they all had the right to participate in the process, even from the beginning. The residents had the power to decide what could and could not happen. She was making the argument that the public defender could help them engage in that process. After she finished speaking, the audience clapped for her.

The meeting continued for over two more hours. People in the audience were given most of the time to make comments and ask questions. The intense arguments continued, only calming down when Andreia spoke and toward the end of the meeting, when everybody was tired. A few townspeople argued that the mining project was halted because of the public defenders, the social movements, and other outsiders. When Andreia had an opportunity to respond, she explained that the public defender was neither for nor against the project. The public defender was there to help ensure that people could participate in the process and that the government and Belo Sun, the company, complied with the conditions set out in the terms of the project that granted the necessary licenses. She explained that after a previous public meeting at a nearby island, some residents had approached her to complain about the mining project. She filed the lawsuit on behalf of those residents. She was working hard to convince the crowd that she was there to help them, no matter how the project moved forward. It seemed as though it was working, at least to some extent.

Andreia's ability to appease the crowd was astounding. The residents in favor of the project had been, and would continue to be, hostile to the other speakers opposed to the dam. It seemed clear to me that Andreia was also opposed to the mining project, and she did not shy away from speaking about the potential negative environmental and social impacts of the mine, which included toxic pollution from chemicals used to extract the gold and displacement, among other effects. During the meeting, though, she focused on the need for people to be actively engaged in the process and the ways her institution could help people do that. She had also built the trust and respect of the community throughout the previous five years. This was important because Andreia knew that she could play a vital role in supporting the townspeople should the mining project go forward, and she used the respect she had developed over time to make her arguments in a way that people would listen to.

This story highlights some of the challenges that social movements and nongovernmental organizations have faced, as well as the ways Brazilian legal institutions have started to play important, if precarious, roles in conflicts over large-scale development projects. Groups like Xingu Vivo and MAB were opposed to big development, infrastructure, and resource-extraction projects for ideological reasons and because these projects tended to have negative consequences for marginalized communities. Furthermore, these activists had witnessed the impacts of Belo Monte and had difficulties accomplishing their goals in that context. Due to their experiences, they believed that the local communities, in order to stop the project, needed to share the sentiment of opposition. The public meeting was their attempt to convince the residents of this village and the surrounding communities that the negative impacts would outweigh any gains. But the meeting showed that many people in the community were emboldened to stand up against these outsiders and strongly support Belo Sun. The activists' views of development differed from those of the people whom they were attempting to support. As active translators and mediators of demands, these activists were doing work that many of the people most directly affected simply did not want.

The public defender stepped into this challenging situation and managed to find a middle ground, despite being clearly aligned with the social movements and being partly responsible for temporarily halting the project. Andreia had arrived with Xingu Vivo, had previously held meetings alongside them, and had no reservations about discussing these partnerships, yet was able to allay fears that she would only work with activists. Toward the end of the meeting, Andreia explained that social movements such as Xingu Vivo have natural partnerships with the public defender because of their shared interest in supporting the community over an assortment of issues, including health care, education, and human rights, more generally. By highlighting the active involvement of the social movements, she was also pleading residents to become involved and stay vigilant in protecting their interests. People once again clapped for her and stated that they were organized better than the people in Altamira before the dam was built. They were ready to fight for their rights and even use the public defenders' services when they felt they needed them. The residents clearly viewed Andreia as a resource, even if she was aligned with the social movements.

The work that the legal institutions had carried out, and were continuing to carry out, was indeed strengthening the legitimacy of those institutions for the future. Belo Sun was fast becoming the next big project that would impact the region in consequential ways. Whether people supported or opposed it, there would undoubtedly be unwanted consequences. Andreia was positioning the public defender as a key ally and mediator.

The work of the public defenders and public prosecutors around the Belo Monte dam and the Belo Sun mining project shows that legal institutions influence claims-making processes in a variety of ways, even in the absence of the courts. As scholarship has shown, they do so by shaping and framing the message of movements and by leveraging legal resources to pressure dominant actors to change their behaviors.[12] They also do so in more direct and surprising ways.

The institutions become leverage points and crucial mediators of claims, as long as those institutions maintain legitimacy in the eyes of both the people they hope to support and the government, companies, or individuals that become the targets of demands. The public defenders and public prosecutors built their legitimacy through creative engagement strategies, which was possible because of their ambiguous positions and the individual lawyers who took it upon themselves to work in support of marginalized communities. Their legitimacy, in turn, enabled these institutions to successfully negotiate for those groups. While this case shows that significant gains can be made through these processes, it also highlights the tenuous nature of this type of work and the barriers that marginalized populations face in receiving fair compensation, even in a context that celebrates democratic processes.

CONCLUSION

Dam construction is a problematic endeavor. The major environmental, economic, and social impacts cannot be easily mitigated. Despite these challenges, few signs point to the abatement of the construction of hydroelectric projects. With their promises of renewable energy, new technology to minimize environmental impacts, and innovative programs to manage the social and economic challenges, dams will continue to be built. Many dams, as well as other large infrastructure projects, will likely be constructed in a way similar to that of Belo Monte. They will be accompanied by investments in local development, new ways for everyday citizens to participate in decision-making processes with government officials, and promises to improve the natural environment and protect the social well-being of residents. As is evident from the stories in this book, these interventions—which reflect democratic developmentalism and represent a step forward in responsible governing—can further complicate life for many people.

The idea of democratic developmentalism—in which the state promotes growth through large-scale infrastructure projects while also working to reduce poverty and increase participatory modes of governance—seems contradictory. Developmental states have historically concentrated power at the federal level. They have ushered in large-scale projects, including dams, with little citizen input or regard for the social or environmental costs. These projects have often led to significant harm for those living nearby but few consequences for those in power. In contrast to the historically authoritarian nature of the developmental model, democracy, at least in theory, means ceding power to communities. It would seem unlikely that a government focused on industrial expansion and infrastructure growth could effectively manage the negative impacts and integrate participatory approaches to governance.[1]

Despite these apparent contradictions of democratic developmentalism, the Brazilian state under presidents Lula and Rousseff, along with other Latin American countries in the early twenty-first century, attempted to join the ideologies of a developmental state with increased democratic practices. Large-scale projects constructed with this approach, such as Belo Monte, allow us to examine the state's ability to accomplish these seemingly conflicting goals. The emergence of the democratic developmental state also elicits questions about the impacts that a new wave of large-scale infrastructure projects has on the population. As the government ushers in mechanisms to reduce the negative impacts, increases opportunities for participation, and provides an abundance of resources to the area, what complications and opportunities arise?

The story of Belo Monte suggests that the democratic developmental approach might, on the one hand, reduce *some* negative impacts and create new opportunities for marginalized populations to voice their claims and make some gains toward better livelihoods. On the other hand, the structural barriers to achieving adequate and just compensation can be extensive. This approach is quite effective at quelling resistance to and creating coalitions of support for such projects, which further complicates claims-making processes.

HYPERDELIBERATION AND THE CO-CONSTRUCTION OF DEMOCRACY

The beginning of the twenty-first century saw the emergence of deliberation and negotiation as a central form of claims making on a global scale. In Brazil, the rise of the progressive political party to power not only expanded resources and increased formal opportunities for more citizens to engage in direct decision-making with one another and government officials, but also helped create the conditions for the widespread belief and celebration of deliberation and negotiation. State-created participatory spaces stand out as the sites of deliberation, and are usually the focus of scholarly work on participatory governance. These types of opportunities emerged in the 1990s in Brazil and became commonplace in multiple institutional settings. In cities and regions across the country, citizens have used government-instituted participatory forums to control parts of their municipal budgets, manage health care offerings, and govern natural resources. These democratic experiments have opened the path for more direct dialogue between people and the government. These spaces have also contributed to the unquestioned expansion of deliberation to make and settle claims.[2]

The conflict over Belo Monte shows us *how* deliberation expanded beyond official spaces to become a ubiquitous aspect of political life. The many stories in this book show how deliberation emerges from top-down process, bottom-up grassroots movements, and through institutional mechanisms, such as legal organizations. In each process and form of engagement, deliberation became the principal feature of claims making, bringing consequences with it. Top-down, state-driven opportunities like CGDEX, the participatory committee that the federal government tasked with instituting sustainable development in the region, divided a previously strong network of allies that once had opposed the dam, all of whom fight for fundamental rights for marginalized populations, including riverine communities, farmers, and the urban poor. These participatory mechanisms also encouraged new kinds of discussions and created somewhat surprising alliances, which offered opportunities for both participants and critics of these initiatives to exert agency and begin projects that could support people in the region.

Processes of deliberation also emerged from the bottom. People impacted by the dam's construction who felt abandoned, excluded, and lacking resources found ways to use new institutions in order to create opportunities for negotiation. In contrast to a state-driven process, the struggles of the fishers and riverine people show us how marginalized groups had to navigate a complex web of institutions and actors in order to gain access to participatory opportunities and capture some of the resources arriving in the region. The well-being and the dignity of the fishers were at stake, so they used protest and emergent organizations to open avenues through which to make demands on the state and Norte Energia. Radical direct action, such as the occupation of the dam site, subsided as quickly as it emerged, and direct negotiation emerged as the primary means through which to make claims. By participating in meetings, events, and negotiations for over five years, the fishers and riverine people learned how to be engaged citizens in participatory processes. The fishers' persistence in engaging in deliberations became their resistance, showing how contestation was moving from "the streets," or in this case the riverbanks, to conference rooms. They forced those in power to recognize them as dam-affected communities and to act on that recognition by providing compensation. In so doing, the fishers were creatively using those spaces to render previously unseen struggles visible and exposing contradictions in the planning and implementation of the programs that accompanied dam construction.

Officials from legal institutions, who became the fishers' closest allies, helped create opportunities for deliberation and became negotiators. The public prosecutors and defenders emerged as primary mediators of claims, but not strictly through the use of the formal legal system. They found their work to be most effective when outside the bounds of the courts. The *threat* of legal action may have helped the public defenders and prosecutors make progress in their causes, but they also came to understand that negotiation helped them progress more efficiently.

Nearly every contentious process and claims-making effort occurred in direct negotiation between the claims makers, state actors, and Norte Energia. The emergence of deliberation across a range of settings shows a deepening of democratic processes in multiple ways. An increasing number of people participated, and the quality of deliberation was such that at least some demands made progress and achieved gains for the marginalized. In addition, both state-empowered participatory governance—which included formal financed opportunities, such as CGDEX—and bottom-up forms of participation had their own kinds of success. Furthermore, as people engaged around Belo Monte, we see how a breadth of mediators, including state institutions, legal organizations, nongovernmental organizations, and even social movement activists, translated claims for dam-affected populations, a sign of strengthening democratic practices. This deepening of democracy was thus co-constructed by state and civil society actors across a range of practices and forms of engagement.

On the other hand, the local democracy they produced was woefully insufficient and overly costly for some participants. Many people were unable to participate in democratic opportunities. Some residents lacked the social, cultural, human, and political capital to engage as equal participants, while others were simply excluded. Even when some marginalized groups managed to gain more voice, they did so *despite* the conditions in which they were making claims, rather than *because* of those conditions. The democratic processes that emerged were slow, tedious, and unequal, requiring participants to be imaginative and persistent. Regular meetings between Norte Energia officials and dam-affected residents might be celebrated as opportunities to engage in direct deliberation, yet negotiations dragged on for months and even years before people saw real progress. Furthermore, processes of deliberation and negotiation required that the legal representatives, nongovernmental organizations, researchers, and community members all maintain legitimacy in the eyes of the officials who held the power to provide compensation. But the legitimacy that each person and organization built—sometimes at great cost and only through creative, time-consuming processes—was tenuous. This all amounted to a twisted form of democratic deepening, but it is the only kind that the people and their advocates could create.

This hyperdeliberation highlights, and is further complicated by, the power of private capital. No longer can citizens—even the people most directly impacted by a project that was conceived of and driven forward by the state—directly engage the government. Instead, though residents affected by the dam might grab the attention and support of state actors, government officials have to find ways to compel a private company to yield to their demands. This complex relationship between those impacted by the dam, their advocates outside of the state apparatus, government officials who might help support them, and a private entity complicates claims-making processes and leads us to interrogate the usually unquestioned positive image of democratic deepening.

DAMS, CLIMATE CHANGE, AND ENVIRONMENTAL JUSTICE

Rising global temperatures are causing irreversible environmental damage, making it pressing to transition to a lower-carbon economy and to find better adaptation strategies. A near consensus of scientists argues that human activity is causing global warming and that the consequent climate change is responsible for more intense and longer storms, increased wildfires, rising sea levels due to polar ice melt, floods, droughts, and other severe weather events. Experts also agree that these impacts will continue to worsen for the foreseeable future.³ While everyone will experience environmental changes in the coming years, the effects of warming global temperatures are not equally distributed. Women, the poor, the Indigenous, and people of color, among others, disproportionately suffer the negative consequences of these changes and have the most limited access to decision-making processes. Given the unequal distribution of the impacts of a changing climate, the issue is not only scientific and technical, but also profoundly social, moral, and ethical.⁴

An environmental justice framework suggests that all people and communities—regardless of race, ethnicity, nationality, gender, age, class, or other social markers—have the right to equal protection from environmental harm and access to environmental decision-making processes. The environmental justice movement has been growing since the 1980s, as residents, scientists, activists, sociologists, and others have thrust stories of environmental inequalities into the spotlight as they struggle for justice. These cases have shown how the benefits and burdens of environmental change—including those associated with climate change—are unequally distributed. Environmental justice organizations have led calls for climate justice, or the ability to address climate change in a just manner that gives all nations and people the opportunity to have equal voice in climate negotiations. The environmental justice movement has also brought attention to the need for a just transition to a lower-carbon economy, a transition that takes politics, justice, access to energy, and historical patterns of development seriously.⁵

The Belo Monte project, with its purported attention to social and environmental sustainability, highlights the challenges in generating energy in an environmentally just manner. As Ed Atkins argues, the conflicts around megadams in the Brazilian Amazon represent debates over what forms of energy production are deemed sustainable and who is able to make those determinations.⁶ Proponents of Belo Monte argued that it would generate clean and needed electricity for Brazil's growing economy and for the millions of people coming out of poverty, while producing few greenhouse gas emissions. Additionally, the dam would be built with a run-of-the-river design, which creates a much smaller reservoir than traditional dams, making it more ecologically friendly. On the one hand, these arguments are worth considering. Brazil's energy demands will continue to increase, and the government could choose energy-production avenues that would be more harmful to the climate and to democracy than dam construction with a democratic developmental approach. On the other hand, a just transition to clean forms of energy would take into consideration the needs and desires of everybody, particularly those most historically marginalized. Merely providing citizens and dam-affected communities access to decision-making once the dam is already being constructed falls exceptionally short of a just or equitable approach to development.

Furthermore, Belo Monte's construction brought both predicted and unforeseen consequences related to the changing climate. Ironically, due to the environmentally friendly design, the impacts of climate change have already contributed to reducing the amount of electricity Belo Monte can produce and threatened the integrity of the dam facility itself. Climate change, deforestation, fires, and land-use choices have led to drier conditions in the region, a trend that scientists predict will continue. These dry conditions affect the dam, because the run-of-the-river design requires enough water flow to ensure electricity production and to maintain a healthy ecosystem both upstream and downstream. Additionally, the engineering of the dam did not fully take into consideration the impacts of low water flow. In October 2019, just before the dam was fully operational, the CEO of Norte Energia sent a letter to the national water agency indicating that a reduced flow of water threatened the structural integrity of the upstream Pimental dam, which was used to divert water to the main power-producing dam. The letter asked for permission to reduce the flow to the energy-producing dam in order to protect the part of Pimental that was at risk. The low water flow also further endangered the already fragile ecosystem and made transportation and fishing on the river even more difficult for residents downstream of the dam. The potential catastrophic failure of the facility served as proof to opponents of the dam who have long argued that Belo Monte—and dam construction more generally—is not a sustainable approach to development or energy production.[7] Indeed, taken together, the problems associated with Belo Monte raise serious doubts about whether such a facility is worth constructing, particularly in the name of mitigating climate change.

While the case of Belo Monte highlights how dams are a fraught means of reducing global warming, the contemporary approach to dam construction nevertheless provides valuable insights into how communities, governments, and other organizations prepare for and respond to climate change. Both dams and global warming lead to human-made environmental changes, usually involve attempts by the state to manage the effects of those changes, and often lead to struggles for environmental justice. As with climate change, dam construction impacts everyone in a region, but leads to particular challenges for marginalized groups. Both climate change and dam construction force people out of their homes, for example, and when this happens, marginalized groups have fewer options. People facing displacement who lack financial resources and social capital are at the behest of whoever manages the mitigation project to address their needs (or not). Without money or connections to people in positions of power, displaced people generally feel obligated to accept whatever compensation or resettlement program is offered to them. These programs, even if they are meant to be helpful, are often deeply flawed, leaving incredible hurdles for people to overcome as their lives are upended. The government approach to resettlement and compensation is thus hugely consequential for those groups.

A close examination of the mitigation programs and conflicts surrounding Belo Monte offers particularly salient lessons for climate change adaptation. On a global scale, the resources to curb the effects of climate change are limited, yet significant and hopefully rising. The Belo Monte project shows us that a meaningful investment into mitigation programs is a necessary but far from sufficient step in dealing with the effects of human-caused environmental harm. The mitigation plans that accompanied dam construction appeared to be progressive. The government invested significant amounts of financial resources into programs, yet we saw that the manner in which the resources were distributed was crucial. Powerful actors could easily co-opt these resources, even in the best-case scenario. These factors altered the social, political, and economic dynamics. While it is difficult to fully prepare for the impacts of these changes, Belo Monte shows the importance of viewing climate-change adaptation through the lens of justice and equality.

MOBILIZING THE STATE, MEDIATING CLAIMS, AND CREATING KNOWLEDGE

When we focus the environmental justice lens on climate change, the Belo Monte case shows us that people concerned with environmental inequality need to ask about, search for, and demand the avenues that will be available for impacted people to make claims and have these claims be answered by those who can provide solutions. In the case of Belo Monte, the most impacted groups without much say in their futures included the fishers, residents of stilt houses, bricklayers, farmers, and Indigenous and riverine communities. Some of these groups managed to overcome the lack of resources and political voice in their struggles for social and environmental justice. In order to navigate the complex social, institutional, and political changes and make gains, these groups mobilized state actors to work on their behalf, found actors to mediate and translate their claims, and shifted understandings by advocating for knowledge that would be seen as legitimate.

The fishers, for example, found avenues through which to open up negotiations and access some of the influx of resources in ways that differed from other struggles for environmental justice. The fishers used protest and acts of resistance to activate, mobilize, and strengthen parts of the state apparatus.[8] The mobilization of institutional actors, such as the Ministry of Fisheries and Aquaculture, the Public Defenders' Office, the Public Prosecutors' Office, and other organizations, expanded the network of advocates for the fishers. These advocates could mediate and translate their claims in the participatory settings the protest had helped them gain access to. In those manufactured spaces of negotiation and deliberation, however, the fishers and riverine people bore the burden of proof. The research carried out by the subcontracted research teams of Norte Energia clashed with the experiences of the people who felt they were being negatively affected by the dam. The fishers and riverine communities had to work with their support networks to shift the terrain of struggle in ways that would alter people's understandings of the dam's impacts and force Norte Energia and government officials to value the anecdotal authority of those affected.

In order to make these advances and build a new form of expertise, the process involved "democratizing and mobilizing science," to use Sabrina McCormick's words.[9] Similar to struggles for environmental justice in other parts of the world, the fishers and riverine communities became integral to the scientific process of establishing research priorities and creating knowledge. In the United States, people who have faced health problems due to the negligence of companies that have dumped chemicals in their neighborhoods have collaborated with experts to show the cause-and-effect relationships. Environmental justice struggles that have been most successful are often those with strong networks and partnerships between citizens and scientists. The residents facing environmental change share their anecdotal stories of disease, displacement, and other harmful effects, while scientists support them through research and expert knowledge. The combination of knowledge and understanding provides legitimacy for the residents' cases in the eyes of the companies and the state.[10]

These citizen-science alliances are crucial, but they are difficult for many communities that lack the necessary resources and capital to create. The case of the fishers and riverine communities in Brazil suggest that communities with lower levels of human, social, and financial capital benefit from mediators within alliances at all stages of claims making. In the case of Belo Monte, officials from legal institutions, as well as nongovernmental organizations, played important intermediary roles. Legal officials—even when not using the legal system—used their positions creatively, to build the legitimacy of the claims that the fishers and riverine people were making. They did so by enrolling experts to carry out rigorous research on the dam-affected communities. The legal agencies were then able to use the research findings as leverage in negotiations with Norte Energia and government officials. This process highlights the importance of mediators translating the experiences and claims of residents into actionable demands. In so doing, the networks of impacted citizens, their mediators, and scientists created critical knowledge to be used in the struggle for justice. This process proved critical because, as other scholars have argued, while most people involved in a conflict or debate use science to make their arguments, they are not all struggling on equal terrain.[11]

This case also shows the importance of mediators allowing affected residents to participate and to remain at the forefront of claims-making processes. In Brazil, these mediating agencies had tenuous legitimacy and authority. The legal officials, for example, could not be seen as overtaking and silencing residents, as they could lose both the trust of those residents and their legitimacy in the eyes of the private and public officials to whom they were making demands. All the advocates of the riverine people made a point to emphasize that the people who lived and fished in and along the Xingu River were the most fully engaged participants. Their claims that affected residents were at the forefront of the struggle were not empty words. The people most impacted by the dam learned how to be engaged, persistent citizens who spoke on their own behalf. These fishers and riverine people never stopped engaging. Throughout their struggles, many of them regularly met with lawyers and nongovernmental organizations, attended and spoke at meetings organized by Norte Energia and government officials, and found other ways through which people would listen to their stories and demands. The people who considered themselves leaders recognized that their ongoing engagement was crucial in order for the group to make progress. For example, when I talked with Jose prior to the meeting of the Working Group of Fishing in December of 2017, he proudly told me that he was one of the most respected fishers and riverine people of the whole group. He explained that he had gained this respect because he had been going to every meeting since the 2012 protest and had always participated in the conversations.

While his story contains much to admire, the experiences of Jose, the fishers, and the network of mediators that worked to support them showcase the problems of the hyperdeliberative moment. These meetings took their emotional and financial tolls—as the participants had to spend their time, money, and energy to attend the events—but he and other regular participants lacked other options. In the end, the resident-centered approach to the struggles for environmental justice led to some positive outcomes, but also took incredible persistence and creativity. Ultimately, the hurdles that these residents had to overcome demonstrate the lack of a structural support system, even in a context of supposedly comprehensive compensation plans.

LAW, DEVELOPMENT, AND STRUGGLES FOR JUSTICE

The public prosecutors and defenders of Brazil are good institutions through which to examine the use of the law in development and the relationship between the law and democratic struggles for social and environmental justice. In the stories throughout this book, the use of law and legal agencies in this relatively remote place emerged during a time of crisis and urgency. This setting provided an opportunity for legal institutions to expand the number of causes that the agencies undertook and to expand the boundaries of their institutions. The officials from these agencies proactively persuaded the people directly impacted by dam construction that it would be worth investing time, energy, and even money to use the law and legal agencies in their struggle for justice. Some of the most marginalized communities who felt the need to change their situations seemed willing to invest in those legal agencies, as there were few other reasonable mechanisms through which to engage.

As the legal agencies strengthened and more residents used their services, the challenges of using the law to make demands became more apparent. The courts are inefficient, cases can be dropped, and the targets of claims, including government and private actors, have financial and political resources to defend themselves. These are significant problems when people are faced with rapid social, political, and environmental changes. In this type of challenging context, the question of *how* the law influences people's political expression becomes even more important.

The law and the legal organizations empowered to "make the law matter," as Lesley McAllister aptly describes the work of the Brazilian public prosecutors, are not one and the same.[12] When the legal system fails, officials can continue their work in other ways, which can be more efficient and effective. In this case, the public prosecutors and the public defenders found creative strategies to navigate the institutional changes occurring during dam construction and to build and use their legitimacy as legal actors to support marginalized populations. They did not discontinue their attempts to use the law to their advantage, but instead used legal processes in conjunction with other mechanisms to push for change. Their outreach strategies helped provide both immediate and long-term support to some communities, as well as strengthen the legitimacy of their institutions as mediators of claims in the eyes of civil society, the state, and dam-affected populations. The case shows us how legal officials can mobilize the law in new ways and make the law, their skills, and their institutions work for them. In so doing, legal agencies become key actors in reconfiguring politics.

The state-run agencies of the public prosecutors' office and the public defenders' office became central actors in these conflicts, in part due to their ambiguously defined positions as government-funded institutions that operate outside, and sometimes against, the state apparatus. This ambiguity allowed the officials to remain independent actors and become respected mediators of claims with all actors involved in the conflict, even when they aligned themselves with activists or dam-affected groups. Given their ability to make claims directly against both Norte Energia and the government, these legal agencies were in a position to have remarkable leverage. They could shape the debates and advocate for the people they aimed to support.

The uniquely powerful position of the public lawyers, who were supportive advocates for many marginalized groups, also came with drawbacks for activists and the people making claims. The power of these agencies was potentially problematic, because some communities came to be too dependent on the officials from legal agencies to support them. For example, in 2017, I asked Jose, one of the riverine people's leaders, how he would ensure that Norte Energia would provide him the monthly stipend that the group of riverine people had recently been promised. With a thankful, yet frustrated tone of voice, he said that Thais Santi, the public prosecutor, would ensure that it came through. He added, "But she is traveling now, so I'll just have to wait. What can I do without her?" Officials like Thais and Andreia Barreto, the public defender, became such integral components in the struggle for rights that many people felt helpless without them. The legal officials were central in shaping the residents' understandings of their situation in ways that empowered previously passive people to be willing to fight for their rights. Jose was one such person who was engaged in the struggle for justice for the first time in his life. He felt emboldened to speak his mind at meetings and earn others' attention. Along with this newfound confidence, however, Jose and others were also reliant on the lawyers. The public prosecutors and defenders relied on the dam-affected people to be at the forefront of the fight, but they, personally, had far less to lose should progress on their cases stall. Given the dependent nature of the relationships, the residents were only partially empowered.

The legal agencies' role as mediators in the struggle for rights is both one of the most promising and one of the most tenuous pathways for change. Individual officials might be incredibly motivated, persistent, and committed to the causes for which they are fighting, but they also work for agencies that can close their offices or move them to other sites. This can leave the people they are supporting in the lurch. Furthermore, lawyers and legal processes are part of the structures that maintain power, control, and the status quo and are thus unlikely to challenge those structures, which more radical groups hope to overtake. While these Brazilian agencies are unique in their mandate and some of the lawyers show a willingness to directly challenge the state, charismatic and reform-minded officials are needed if the law and legal institutions are to assume prominent roles in struggles for justice.

WOMEN AND THE STRUGGLE FOR JUSTICE

To not explicitly recognize the role women have played in the struggles for rights, basic needs, and compensation before and after the construction of Belo Monte would be doing a disservice to this story of resistance and collective action. The women whose stories form the backbone of this book are not only remarkable in their own right, but they are also in the company of many other women who have been at the forefront of the global environmental movement, particularly the environmental justice movement, for over a century.

The role of women in the mainstream environmental movement in the United States in the late 1800s and early 1900s often went unnoticed, even as they made influential interventions that altered understandings of the natural world. While middle- and upper-class white men took on most of the leadership roles in the increasingly professionalized movement, women led environmental crusades to reduce pollution and improve urban environments. In the middle of the century, the role of women in the environmental movement began gaining more attention when authors Marjory Stoneman Douglas and Rachel Carson upended perspectives of how people related to the natural world. Researchers like Dr. Jane Goodall and activists such as Julia Butterfly Hill continued to elevate ecological awareness and strengthened environmentalism in the United States. Despite these and a few other noteworthy figures, men have continued to outnumber women in participation and leadership positions in mainstream environmental movements.[13]

Struggles for environmental justice, in contrast to mainstream environmentalism, have been predominantly led and organized by women. Lois Gibbs's leadership in raising public attention to and seeking compensation for the toxic waste crisis in Love Canal, New York, led to the creation of the Superfund Act. Women in Appalachia organized an environmental justice movement to protect regional communities from the dangers associated with mountaintop-removal coal mining. Black women, such as Peggy Shepard, Majora Carter, and Karen Washington, among many others, have been pioneers in addressing environmental racism since the 1980s. Winona LaDuke and other Indigenous women have become the faces of environmental justice fights on Indigenous lands, of regional responses to disasters, and during Indigenous protests at the United Nations' climate change negotiations. Globally, women like Kenyan-born professor Wangari Maathai and Indian environmentalist Vandana Shiva, among many others, have also charted the course for environmental conservation and struggles for food sovereignty. Countless other women have led movements from the local to transnational levels.

Scholarship suggests that the traditional identities women have often held help explain why they outnumber men in environmental justice struggles. Women have used their identities as "resources of resistance," to borrow from Celene Krauss.[14] For example, many women have drawn on their roles as mothers as they defend their families and communities. Additionally, women's activism has often been rooted in local issues that they know well and experience firsthand. As a result, even the female activists who have little scientific training or political experience prior to their engagement are able to reimagine what it means to have expertise and how knowledge is constructed. How these women claim knowledge and authority—and therefore experience activism—is informed by their gender.[15]

Some of these same global dynamics have played out along the Transamazon. Women have been at the forefront of struggles for justice in the region for decades. Regional activism, in a broader sense, has been informed by the experiences of women. Since the military dictatorship began colonizing the region in the 1970s, women have mobilized to fight for gender equality and address the high rates of gendered violence. Most cases of violence against women went unreported, and few perpetrators were ever prosecuted. In response, women organized, founded the Movement of Women Workers of Altamira from the Country and the City, and were successful in creating a mechanism through which women could safely file claims against domestic abuse. Women—informed by, but not restricted to, their experiences with violence—also led struggles on issues that went far beyond those related specifically to gender. They took leadership roles in organizing communities, resisting dam construction, and pressuring the state and federal government to provide basic infrastructure. Women used the cultural and organizational power of the Catholic Church—and eventually the political vehicle of the Workers' Party—to help organize residents in both rural and urban areas, leading communities in discussions over pressing concerns, their rights, and how they could go about fighting for those rights. Through these efforts, women were crucial for improving education, health care, and basic infrastructure throughout the region. Irá, who was a leader of the anti-dam group Xingu Vivo and a founder of the regional Black Movement, once proclaimed when describing the previous three decades, "The government takes credit for the improvement of life along the Transamazon, but it was really the women behind everything. We fought for the environment, for rural families, for schools, and health care. The women behind these initiatives were really the people who accomplished things in this region!"

Some women have paid dearly for their activist work in the region. Dorothy Stang, an American-born nun who dedicated her life to defending the environment and the rural poor in Brazil, was known for helping small-scale farmers use natural resources without causing deforestation and defending those farmers from ranchers who tried to take their land. Her actions made her the target of death threats, and in 2005, a hired gunman killed her in Anapu, a town not far from Altamira. Most of the women whose stories are in this book were deeply affected by the assassination, as they had worked closely with Stang, but her death compelled them to carry on their work.

Women continued to be at the forefront of the struggle for justice while the dam was being built. Most of the women who had fought for rights and had worked to block dam construction when it was first introduced in the 1980s remained active in advocating for justice during construction, even as they used different avenues through which to engage. Some continued to protest Belo Monte, while others worked to support of people who would be impacted. Antonia Melo and Irá, for example, had helped start the women's movement, marched in the streets against the first dam proposals, and eventually formed Xingu Vivo. They remained the staunchest opponents to Belo Monte, supported the fishers during their protest, and were a constant presence in the conflict over the dam. They were not shy about their feminist perspectives and struggles for gender equity, and this informed their activism. In fact, they occasionally critiqued other organizations for not paying direct attention to the challenges that women faced in the more remote communities in the region. Similarly, Simone and other women from FVPP, the foundation that supports regional small-scale farmers, had also been leaders of the women's movement and the church's efforts to organize rural communities. During dam construction, they discontinued much of their anti-dam activism, but remained leading voices for gender equality and vocal participants in CGDEX and other efforts to bring development to the region. Eliana, another longtime resident and activist in the region, led the successful campaign for the creation of a technical group on health in CGDEX and never shied away from speaking out against what she viewed as injustice. She also became a leading voice in addressing the problems, such as the lack of water and high rates of violence, in the Urban Resettlement Community where she lived.

Furthermore, most of the people who arrived in the region to support communities affected by Belo Monte during construction were women. Andreia was the most active and vocal public defender to work with displaced communities and others impacted by construction. Likewise, Thais, the federal public prosecutor, took it upon herself to provide vital assistance to the fishers, riverine communities, and Indigenous people. Together, these officials from legal agencies, along with female representatives of federal government agencies, found creative ways to work together to support groups like the fishers, riverine people, and others who would lose their homes or businesses. The "gang of women," as Andreia referred to these informal alliances, showed the degree to which women, in particular, came together to confront the powerful forces behind the construction of the dam and the harm it was causing marginalized populations. Together, these stories show that a remarkable number of women were at the forefront of the struggle for justice as Belo Monte was being built. Women have most often been the ones to lead the struggle, to take the difficult positions, and to stand up for themselves and the people most disenfranchised and harmed by Belo Monte.

THE FUTURE OF DEMOCRACY IN BRAZIL

In recent years, we have often heard that democracy is under threat around the world. The rise of far-right groups that celebrate dictators and authoritarian leaders, the spread of "fake news," the erosion of legal institutions, the diminishing belief in journalism, and the delegitimization of science, among other factors, all point to a dwindling of democratic functioning and a risk to the foundations of the democratic order. The institutions that hold democracies together, like an independent press and a nonpartisan fair judicial system, have weakened in many places. With corporate power increasing, ordinary people find it more difficult to hold large companies responsible and in check. These and other threats have repeatedly made Brazilian democracy precarious—and not only in the twenty-first century but for more than a hundred years. The country has waffled between democratic and authoritarian leadership, and many Brazilians remember well life under the military dictatorship. When the economy weakens or people tire of corrupt politics, the distrust of government institutions rises, and the possibility of a return to authoritarian rule grows.

Proponents of the democratic order in Brazil are anxious about its ability to survive. The conflict over Belo Monte suggests that some of these fears are warranted, even though the project began under the Workers' Party government. Many people were officially excluded from formalized opportunities to participate in decision-making processes, and the central government unilaterally decided to move forward with the project even with significant opposition. Additionally, the project was implicated in Lava Jato, the largest corruption scandal in Brazil's history. Few would suggest that the government instituted Belo Monte in a democratic way; instead, the dam manifested the power and control of a developmental state and the private sector. These powerful actors failed to take into account the full needs of the people who would be most affected by dam construction. Furthermore, local networks of people who had once been united in their struggles for justice fractured just as the local populations were feeling the brunt of the effects of dam construction.

Despite the undemocratic and destructive nature of Belo Monte, the stories in this book suggest that the future of democracy is not all grim. People at the local level, given the right institutional structures, can manage to open and deepen avenues for democratic participation, even under challenging conditions. The policies and programs that accompanied dam construction created particular challenges for claims-making processes, yet people still demanded democratic practices and used the institutions at their disposal to voice their concerns and needs in meaningful ways. The fishers and riverine communities, against all odds, managed to make gains in their efforts to receive fair and adequate compensation. Similarly, the participants of state-created participatory forums forced the space to be used in the most equitable way possible, which allowed new voices to enter the conversation and helped create new and surprising alliances that proved helpful in claims-making processes. Even though President Bolsonaro shut down CGDEX in 2019, people had used the space in ways that would undoubtedly have lasting impacts.

In many ways, the actual functioning of democracy matters most, and can be most effective, at the local level. It is in localized situations that we can examine whether, how, and how much everyday people and the most marginalized communities are able to participate in their own governance. By paying attention to local dynamics, we can also see the extent to which communities and those in positions of power respond to the claims, demands, and needs of residents. Furthermore, we are able to witness the strength, legitimacy, and authority of the institutions that are central to a democracy. As people struggle to find their voice, they use the institutions and organizations that are vital for democracies to thrive. The work that people put into using and strengthening these institutions locally can, in turn, strengthen democracy more widely. When the fishers and riverine communities created spaces of negotiation, they were producing democratic practices. The public defenders and public prosecutors were not only becoming key mediators for marginalized groups, but also bolstering their authority and expanding and enforcing the roles that their agencies could play. When the participants of CGDEX engaged in the forum, they were also creating democratic practices. All of these engaged residents were pushing back against the threats to their own lives and those they supported, but also to the factors that threaten democracy. By creating opportunities for deliberation, activating state mechanisms to work on their behalf, and using institutions in innovative ways to garner support and encourage participation, people were strengthening democracy. The local-level bolstering of institutions central to democracy provides reason to believe that democracy can be deepened even in difficult situations.

In addition to providing insights into how democracy emerges, these stories also show *who* creates and supports the strengthening of meaningful democratic practices. Everyone—from staunch anti-dam activists to proponents of the project, from impacted communities to government agencies, and from lawyers to business people—supported and engaged the institutions and mechanisms that allow democracy to thrive. Democracy, in its most consequential forms, is never one-sided. It is co-constituted, at once instituted by those in power and constructed through grassroots efforts, radical action, and protest. Democracy is produced, particularly at the local level, when actors from various sectors work together and in conflict.

The local democratic practices we see in this book did not emerge in a vacuum, but rather reflect and contribute to the changing dynamics of political conflict in Brazil and beyond. The rise of the left had the paradoxical effect of weakening traditional forms of claims making, giving rise to other avenues through which to make demands. Direct acts of protest have not disappeared, but they have dwindled in frequency and have become a less significant aspect of forcing change than in the past. Meanwhile, new participatory opportunities, both those created by the government and also those forced from below, have emerged, supported by a hyperdeliberative moment in which participation and deliberation are celebrated by everyone. The ubiquitous nature of deliberation necessitates strong mediators and translators of claims, and the organizations that implement the law have begun filling these roles. In so doing, legal agencies and the officials in those agencies are becoming central to conflicts over rights, recognition, and compensation for harm to marginalized communities. The stories in the book show that the public lawyers that embrace social justice perspectives and creative approaches to their work are emerging as crucial actors in the functioning of democracy and the ability of people to have their voices heard, recognized, and answered. For generations, the rule of law has been a key indicator of the strength of a democracy. Here, we see that it is not just the rule of law that is a crucial institution, but also the legitimacy and authority of legal officials who serve as effective mediators of claims within and beyond the confines of the courtroom and legal proceedings.

If we understand democracy to mean, at least in part, that more people are able to make claims on government that produce tangible results, this book highlights the uneven, tenuous nature of democracy. Throughout Brazil, as in the case of Belo Monte, an increase in participatory opportunities has incorporated more voices in decision-making processes. These deliberative forms of engagement have been slow, have failed to fully satisfy any of the parties involved, and have risked replicating power imbalances between marginalized populations, the state, private companies, and other actors. Nevertheless, the people involved often refuse to settle for inadequate and unjust processes. The onus to make deliberation inclusive and effective has fallen on the most marginalized communities and the people and institutions who serve as the mediators and translators of their claims. This book showcases these and the many other challenges that marginalized populations face in the contemporary moment and the burdens these communities endure to maintain and strengthen democracy. Even as the democratic order may be in jeopardy, the people who struggle for justice and for their own lives also represent and enact the hope that meaningful democracy will endure.

ACKNOWLEDGMENTS

The simple act of saying "thank you" and "obrigado" does not do justice to the many contributions of so many people toward this book. Nevertheless, I offer my notes of appreciation here.

My utmost thanks go to the people who opened their homes, minds, and hearts to me in Brazil. Whether or not their stories are in these pages, I have learned a great deal about life and struggle from hundreds of people, particularly in and around Altamira, who gave me their time and shared their thoughts about issues central to their lives. Many of these people faced imminent threats to their livelihoods, yet still spent time, talked, and laughed with me.

The academic community in Brazil has been vital to my work. My thanks go to many colleagues whom I know through the Center for Population Studies (NEPO) at UNICAMP. Álvaro de Oliveira D'Antona, Thais Tartalha do Nascimento Lombardi, Julia Côrtes, Igor Cavallini Johansen, Roberto do Carmo, Ricardo Dagnino, Gilvan Guedes, Ana Claudia Teixeira, and a team of other faculty, graduate students, and undergraduates patiently helped with my Portuguese when I first arrived, introduced me to the Amazon and São Paulo, and showed me what Brazilian hospitality is all about. They have continued to be trusted collaborators and friends. A special thanks to Thais and Julia, as well as their partners, Emerson Lombardi and Henrique Santos, who graciously welcomed me into their homes.

I am also grateful to those at the Altamira campus of UFPA. José Herrera provided logistical support and indispensable perspectives on Belo Monte and life in the region. His research and help, along with that of colleagues Daniel Vallerius, Luiz Fernando Roscoche, and many other faculty and students, was incredibly important. Others in Altamira supported me in many different ways. Alessandra Moura, Elaine Moura, and Monica Roche, along with many of their families and friends, taught me about the social and political dynamics in Altamira and made my life there more enjoyable. I am also very thankful to Douglas Tyminiak. He and his family opened their home to me many times, and he never shied away from guiding me on my fieldwork adventures.

I am forever grateful to those in Complexo do Alemão, from where I wrote much of my dissertation. I thank Irailda, Janio, Joel, and Marilia for sharing their home. Irene, Francisco, Bahiana, Maria, Paulo, and many others made the community feel safe and welcoming. I also thank Henrique Gomez da Silva and Andreza Jorge, activists and residents of Complexo da Maré, for their support and friendship for me and my family.

Likewise, the support I have received in the United States is extensive. It is difficult to adequately show appreciation to my two dissertation advisors, Leah VanWey and Gianpaolo Baiocchi, both of whom continue to be mentors and friends. Leah took a chance with me when I was not sure what my path in the academic world would be. She introduced me to the Brazilian Amazon and guided me in a world of research about which I knew little. Gianpaolo provided valuable insights into ethnography and Brazil's experiments with participatory democracy. He also encouraged me to think outside the traditional boundaries of academia, and our other research together inspired and motivated me in many ways. Together, Leah and Gianpaolo made my work more interesting and exciting. I thank them not only for their support in the research and writing of what would eventually become this book, but also for showing me what it means to be a researcher and scholar with integrity. Most importantly, their belief and trust in me never seemed to waiver.

Many others at Brown University supported me and my work. J. Timmons Roberts and Patrick Heller not only gave me useful comments on my dissertation but also provided inspiration and guidance throughout my time at Brown. I am also grateful to my co-authors of *The Civic Imagination*: Elizabeth Bennett, Alissa Cordner, Stephanie Savell, and Gianpaolo Baiocchi. We learned how to research and write together, and the themes and ideas we developed have informed my teaching and other scholarship. My appreciation also goes to Heather Randell, who has been a terrific research partner in the field, co-author, colleague, and friend. I am fortunate to have entered graduate school with five terrific women—Alissa Cordner, Erica Mullen, Mim Plavin, Trina Vithayathil, and Mujun Zhou—who supported me through challenges and celebrated accomplishments with me. I thank the many other graduate students with whom I sat around tables, in and out of the classroom, debating and discussing ideas and from whom I learned so much. And I thank Joan Picard, Muriel Bessette, Karl Dominey, Kristen Soule, Amanda Figgins, Shane Martin, and the others who supported Brown's sociology program and made everything function smoothly behind the scenes.

Thank you to the many colleagues at Bard who give the college a rich intellectual life and make it an enjoyable place to grow as a scholar, teacher, and community member. A particular note of appreciation goes to the sociology program, including Allison McKim, Laura Ford, and Yuval Elmelech, who have shown me a great deal of support and whose comments about my work have helped advance it. The EUS program—including various iterations of the steering committee and the wider EUS community—has provided an ideal interdisciplinary home for my research, teaching, and community work. I also appreciate everyone who attended my faculty seminar and those who discussed my work with me in other settings, as it deepened the analysis. Thank you to Melissa Germano, who has helped with day-to-day life at the college. It has also been a joy to work with the students at Bard.

Many scholars and friends outside of my intellectual homes have inspired and helped me along the way. I thank Rebecca Abers, Claudio Benzecry, Grant Burrier, Salo Coslovsky, Simone Gomes, Gabriel Hetland, Peter Houtzager, Philip Lewin, Ann Mische, Keith Morton, Marianna Olinger, Anthony Pahnke, Ana Pereira, and Rebecca Tarlau. A special note of thanks to Pablo Lapegna and Kathryn Hochstetler for their thoughtful, detailed reviews of this manuscript. This manuscript also benefitted from the editors and anonymous reviewers at the *Journal of Peasant Studies* and *Society and Natural Resources*, who provided comments on articles I published that use some of the research and analysis that appear this book.

A very special thank-you goes to Charles Strozier, who helped bring the book's stories to life and gave me the confidence to carry on.

Thank you to several institutions that generously funded and supported this project at various stages, including the National Science Foundation, the Fulbright U.S. Student Program, the Center for Latin and Caribbean Studies and the Sociology Department at Brown University, the São Paulo Research Foundation (FAPESP), the Political Science Institute at the University of Brasília, the State University of Campinas (UNICAMP), and Bard College. I also thank the Brazil section of the Latin America Studies Association for the support and recognition of my dissertation.

It has been a pleasure working with Peter Mickulas and the team at Rutgers University Press, who did excellent work in bringing this book to press.

Finally, I thank my parents, Toni and George, as well as my sister, Heather, for their unconditional love and for providing me with the tools to be successful in whatever path I chose. My parents' commitment to education and support in every way has made this book and my career possible. My extended family—especially my cousins Jeff and Shane—have been an important source of support. I am fortunate to have married into a family who have given me time, energy, love, and encouragement to finish this book. Nat, Cheen, and John have welcomed me into the family and regularly make me laugh. My mother-in-law, Catherine Savell, has not only provided childcare at crucial moments, but has also inspired me with her excellent university teaching and devoted community work, both domestically and abroad. I appreciate the love and acceptance of my late father-in-law, Geoff Savell, and I miss our spirited debates. I like to think that he would have enjoyed reading this book and that he would have spent hours asking me questions about the big arguments and the smallest details.

Stephanie, Sabine, and Isabelle have been incredibly patient and loving with me during the many hours, days, months, and years that I have been working on this project. Without Stephanie, I would not have finished—and maybe not even started—this book. From the beginning, when she was the perfect companion during portions of my fieldwork, to the end, when she provided important editing work, she has helped make this a better book than it would have been. She is a partner in every sense of the word. I thank her for all of this and much more, for showing me new ways of looking at the world, and for believing in me.

LIST OF ABBREVIATIONS

BNDES	Banco Nacional de Desenvolvimento Econômico e Social	National Bank for Economic and Social Development
CCBM	Consórcio Construtor Belo Monte	Belo Monte Construction Consortium
CEB	Communidade Eclesiais de Base	Christian Base Community
CGDEX	Comitê Gestor do Plano de Desenvolvimento Regional Sustentável do Xingu	Steering Committee of the Regional Sustainable Development Plan of the Xingu
CIMI	Conselho Indigenista Missionário	Indigenous Missionary Council
CNBB	Conferência Nacional dos Bispos do Brasil	National Conference of Brazilian Bishops
CNEC	Consórcio Nacional de Engenheiros Consultores	National Consortium of Engineering Consultants
CRAB	Comissão Regional de Atingidos por Barragens	Regional Commission of People Affected by Dams
DNAEE	Departamento Nacional de Águas e Energia Elétrica	National Department of Water and Electrical Energy
DPE	Defensoria Pública do Estado	Public Defenders' Office of the State
DPU	Defensoria Pública da União	Public Defenders' Office of the Union
EIA		Environmental impact assessment
FUNAI	Fundação Nacional do Índio	National Indian Foundation
FVPP	Fundação Viver, Produzir e Preservar	Living, Producing, and Preserving Foundation
Ibama	Instituto Brasileiro do Meio Ambiente e dos Recursos Naturais Renováveis	Brazilian Institute of the Environment and Renewable Natural Resources
ISA	Instituto Socioambiental	Socio-environmental Institute
MAB	Movimento dos Atingidos por Barragens	Movement of People Affected by Dams
MDTX	Movimento Pelo Desenvolvimento da Transamazônica e Xingu	Movement for the Development of the Transamazon and Xingu
MMTA-CC	Movimento de Mulheres Trabalhadoras de Altamira—Campo e Cidade	Movement of Women Workers of Altamira—Country and City

MPA	Ministério da Pesca e Aquicultura	Ministry of Fisheries and Aquaculture
MPF	Ministério Público Federal	Federal Public Prosecutors' Office
MPST	Movimento Pela Sobrevivência na Transamazônica	Movement for the Survival on the Transamazon
PAC	Programa de Aceleração do Crescimento	Growth Acceleration Program
PAS	Plano Amazônia Sustentável	Sustainable Amazon Plan
PBA	*Projeto Básico Ambiental*	*Basic Environmental Project*
PDRSX	*Plano de Desenvolvimento Regional Sustentável do Xingu*	*Regional Sustainable Development Plan of the Xingu*
PIN	Programa de Integração Nacional	National Integration Plan
PT	Partido dos Trabalhadores	Workers' Party
RIMA	*Relatório de Impacto Ambiental*	*Summary of the Environmental Impact*
RUC	Reassentamento Urbano Coletivo	Urban Resettlement Collective
SBPC	Sociedade Brasileira para o Progresso da Ciência	Brazilian Society for the Advancement of Science
STTR	Sindicato de Trabalhadores e Trabalhadoras Rurais	Rural Workers' Union
UFPA	Universidade Federal do Pará	Federal University of Pará
Xingu Vivo	Movimento Xingu Vivo Para Sempre	Xingu Forever Alive Movement

NOTES

INTRODUCTION

1. My usage of the term "democratic developmentalism" originates from Mark Robinson and Gordon White, eds., *The Democratic Developmental State: Politics and Institutional Design, Oxford Studies in Democratization* (New York: Oxford University Press, 1998). Other scholars and I suggest the concept is helpful to describe Brazil's approach to development from 2003 to 2016. See, for example, Grant Burrier, "The Developmental State, Civil Society, and Hydroelectric Politics in Brazil," *Journal of Environment and Development* 25, no. 3 (2016): 1–27; Kathryn Hochstetler and Ricardo J. Tranjan, "Environment and Consultation in the Brazilian Democratic Developmental State," *Comparative Politics* 48, no. 4 (July 1, 2016): 497–516; Peter Taylor Klein, "Engaging the Brazilian State: The Belo Monte Dam and the Struggle for Political Voice," *Journal of Peasant Studies* 42, no. 6 (March 20, 2015): 1–20.
2. This definition of deep democracy borrows from Arjun Appadurai and John Gaventa. Arjun Appadurai, "Deep Democracy: Urban Governmentality and the Horizon of Politics," *Environment and Urbanization* 13, no. 2 (2001): 23–43; John Gaventa, "Triumph, Deficit or Contestation? Deepening the 'Deepening Democracy' Debate," *IDS Working Paper*, no. 264 (July 2006): 1–34.
3. Operation Lava Jato (Car Wash) was a criminal investigation into money laundering and corruption in Brazil. Soon after the investigation began in 2014, it became Latin America's largest corruption scandal. It implicated the state-run petroleum company Petrobras, other major Brazilian construction firms, politicians at all levels of government from many different parties, and former presidents, including Luiz Inácio Lula da Silva, who was sent to prison but later released. The operation investigated and uncovered bribery schemes in which contractors paid government officials and politicians in order to receive lucrative contracts. In 2018, the federal police launched a phase of the investigation that focuses on bribery payments made by the builders of Belo Monte to win the contract for the project.
4. A growing body of literature examines recent regulatory and development practices of leftist governments in Latin America, how their policies resemble the neoliberal agenda of previous decades, and the conflicts that emerge between grassroots activists on the left and the left of center governments. See Derrick Hindrey, *From Enron to Evo: Pipeline Politics, Global Environmentalism, and Indigenous Rights in Bolivia* (Tucson: University of Arizona Press, 2013); Thea Riofrancos, *Resource Radicals: From Petro-Nationalism to Post-Extractivism in Ecuador* (Durham, NC: Duke University Press, 2020); Kregg Hetherington, *The Government of Beans: Regulating Life in the Age of Monocrops* (Durham, NC: Duke University Press, 2020).
5. For examples of scholarship on mobilization, demobilization, resistance, and accommodation around development projects, see Stephanie A. Malin, *The Price of Nuclear Power: Uranium Communities and Environmental Justice* (New Brunswick, NJ: Rutgers University Press, 2015); Pablo Lapegna, *Soybeans and Power: Genetically Modified Crops, Environmental Politics, and Social Movements in Argentina* (New York: Oxford University Press, 2016).
6. While much of the literature on social movements, collective action, and environmental and social justice tends to focus on how communities respond when faced with a scarcity of resources and dearth of opportunities, there are noteworthy exceptions. Samer Alatout and Jessica Barnes, for example, show the importance of examining the everyday politics associated with the abundance of water. Samer Alatout, "Bringing Abundance into Environmental

Politics: Constructing a Zionist Network of Water Abundance, Immigration, and Colonization," *Social Studies of Science* 39, no. 3 (2009): 363–94; Jessica Barnes, *Cultivating the Nile: The Everyday Politics of Water in Egypt* (Durham, NC: Duke University Press, 2014). Likewise, Cynthia Simmons has argued that a combination of resource scarcity and abundance, along with the presences of active social movements, has interacted to create violent outbreaks in the Brazilian Amazon. Cynthia S. Simmons, "The Political Economy of Land Conflict in the Eastern Brazilian Amazon," *Annals of the Association of American Geographers* 94, no. 1 (2004): 183–206. These examples, and much of the other literature on the impacts of abundance on collective action, have focused on the presence and discovery of natural resources, rather than the influx of social, institutional, and economic resources into an area.

7. Throughout the book, I use "Norte Energia representatives" to refer to officials employed directly by Norte Energia, which is the consortium of public and private entities developed with the sole purpose of constructing Belo Monte, and to employees of companies that Norte Energia has subcontracted to carry out specific construction and mitigation programs.

8. I attempted to follow two methodological approaches from actor-network theory: agnosticism and symmetry. Agnosticism refers to approaching a controversy and the actors involved in that controversy with impartiality. This approach includes taking care to avoid prematurely judging how people assess and analyze the social world and situations around them. Symmetry means using the same analytical framework for all of the actors involved. These approaches motivated me to engage with a wide range of people and organizations who had different, often conflicting, views on the dam. For more on using agnosticism and symmetry, see Michel Callon, "Some Elements of a Sociology of Translation: Domestication of the Scallops and the Fishermen of St. Brieuc Bay," in *Power, Action and Belief: A New Sociology of Knowledge?*, ed. J. Law (London: Routledge, 1986), 196–233; David Bloor, *Knowledge and Social Imagery*, 2nd ed. (Chicago: University of Chicago Press, 1991); Gianpaolo Baiocchi, Elizabeth A. Bennett, Alissa Cordner, Peter Taylor Klein, and Stephanie Savell, *The Civic Imagination: Making a Difference in American Political Life* (Boulder, CO: Paradigm Publishers, 2014).

9. When appropriate and with permission, I audio recorded formal interviews. Some respondents declined to be recorded. In most cases, I took handwritten notes during interviews, regardless of whether I was recording or not.

10. Malin, *The Price of Nuclear Power*, 9.

11. Kathryn Hochstetler, "The Politics of Environmental Licensing: Energy Projects of the Past and Future in Brazil," *Studies in Comparative International Development* 46 (October 5, 2011): 349–71.

12. For more on the relational sociological approach to examining civil society, processes of democratic deepening, and how public and political life are maintained and changed, see Margaret R. Somers, "Citizenship and the Place of the Public Sphere: Law, Community, and Political Culture in the Transition to Democracy," *American Sociological Review* 58, no. October (1993): 587–620; Mustafa Emirbayer, "A Manifesto for a Relational Sociology," *American Journal of Sociology* 103, no. 2 (1997): 281–317; Gianpaolo Baiocchi, *Militants and Citizens: The Politics of Participatory Democracy in Porto Alegre* (Stanford, CA: Stanford University Press, 2005); Gianpaolo Baiocchi, Patrick Heller, and Marcelo Silva, *Bootstrapping Democracy: Transforming Local Governance and Civil Society in Brazil* (Stanford, CA: Stanford University Press, 2011). For a "strategic-relational" approach as it applies specifically to the law and politics, see Christos Boukalas, "Politics as Legal Action/Lawyers as Political Actors: Towards a Reconceptualisation of Cause Lawyering," *Social and Legal Studies* 22, no. 3 (September 11, 2013): 395–420. For details of how W. E. B. Du Bois developed the relational methodological approach, see José Itzigsohn and Karida L. Brown, *The Sociology of W. E. B. Du Bois: Racialized Modernity and the Global Color Line* (New York: New York University Press, 2020).

13. Hans Joas, *Pragmatism and Social Theory* (Chicago: University of Chicago Press, 1993), 4.
14. The American pragmatist school of thought refers to the work of George Hebert Mead, John Dewey, and Charles Sanders around the turn of the twentieth century. For examples of pragmatism, see George Herbert Mead, "The Genesis of the Self and Social Control," *International Journal of Ethics* 35, no. 3 (1925): 251–77; John Dewey, *Human Nature and Conduct: An Introduction to Social Psychology* (New York: Henry Holt, 1922). The work of the American pragmatists has seen a resurgence in sociological literature, particularly in scholarship examining theoretical ideas of agency and studies that pay attention to the role of future-oriented thought on present-day action. See Joas, *Pragmatism and Social Theory*; Mustafa Emirbayer and Ann Mische, "What Is Agency?," *American Journal of Sociology* 103, no. 4 (1998): 962–1023; Neil Gross, "A Pragmatist Theory of Social Mechanisms," *American Sociological Review* 74, no. 3 (June 1, 2009): 358–79; Ann Mische, *Partisan Publics: Communication and Contention across Brazilian Youth Activist Networks* (Princeton, NJ: Princeton University Press, 2009); Margaret Frye, "Bright Futures in Malawi's New Dawn: Educational Aspirations as Assertions of Identity," *American Journal of Sociology* 117, no. 6 (May 22, 2012): 1565–624. For another example of scholarship that bridges relational sociology with pragmatism, see David Smilde, *Reason to Believe: Cultural Agency in Latin American Evangelism* (Berkeley: University of California Press, 2007).
15. Marco Aurelio dos Santos, Luiz Pinguelli Rosa, Bohdan Sikar, Elizabeth Sikar, and Ednaldo Oliveira dos Santos, "Gross Greenhouse Gas Fluxes from Hydro-Power Reservoir Compared to Thermo-Power Plants," *Energy Policy* 34 (2006): 481–88; Nathan Barros, Jonathan J. Cole, Lars J. Travnik, Yves T. Prairie, David Bastviken, Vera L. M. Huszar, Paul del Giorgio, and Fábio Roland, "Carbon Emission from Hydroelectric Reservoirs Linked to Reservoir Age and Latitude," *Nature Geoscience* 4 (2011): 593–96; Bridget R. Deemer, John A. Harrison, Siyue Li, Jake L. Beaulieu, Tonya DelSontro, Nathan Barros, José F. Bezerra-Neto, Stephen M. Powers, Marco A. dos Santos, and J. Arie Vonk, "Greenhouse Gas Emissions from Reservoir Water Surfaces: A New Global Synthesis," *BioScience* 66, no. 11 (November 1, 2016): 949–64.
16. The Intergovernmental Panel on Climate Change (IPCC) has stated that while it is difficult to estimate with any certainty where and to what extent displacement and migration will occur as a result of climate change, studies show that changing weather patterns will likely lead to significant movement of people, particularly from and to agriculture-dependent countries, tropical regions, and small islands. The IPCC has covered these impacts in various reports. For one example with specific details and citations to scientific studies, see section 3.4.10.2 on migration, displacement, and conflict in O. Hoegh-Guldberg, D. Jacob, M. Taylor, M. Bindi, S. Brown, I. Camilloni, A. Diedhiou, R. Djalante, K. L. Ebi. F. Engelbrecht, J. Guiot, Y. Hijioa, S. Mehrotra, A. Payne, S. I. Seneviratne, A. Thomas, R. Warren, and G. Zhou, "Impacts of 1.5°C Global Warming on Natural and Human Systems," in *Global Warming of 1.5°C. An IPCC Special Report on the Impacts of Global Warming of 1.5°C above Pre-industrial Levels and Related Global Greenhouse Gas Emission Pathways, in the Context of Strengthening the Global Response to the Threat of Climate Change, Sustainable Development, and Efforts to Eradicate Poverty*, ed. V. Masson-Delmotte, P. Zhai, H.-O. Pörtner, D. Roberts, J. Skea, P. R. Shukla, A. Pirani, W. Moufouma-Okia, C. Péan, R. Pidcock, S. Connors, J.B.R. Matthews, Y. Chen, X, Zhou, M. I. Gomis, E. Lonnoy, T. Maycock, M. Tignor, and T. Waterfield, https://www.ipcc.ch/site/assets/uploads/sites/2/2019/05/SR15_Citation.pdf. 2018.
17. World Commission on Dams, *Dams and Development: A New Framework for Decision-Making* (London: Earthscan, 2000).
18. For an example of research that shows an increase in livelihood outcomes after displacement from dam construction, see Heather Randell, "The Short-Term Impacts of Development-Induced Displacement on Wealth and Subjective Well-Being in the Brazilian Amazon," *World*

Development 87 (November 1, 2016): 385–400. For examples of negative effects of displacement and some recommendations for mitigating such effects from development-induced displacement, see Chris McDowell, ed., *Understanding Impoverishment: The Consequences of Development-Induced Displacement* (Providence, RI: Berghahn Books, 1996); Michael M. Cernea and Julie Koppel Maldonado, eds., *Challenging the Prevailing Paradigm of Displacement and Resettlement: Risks, Impoverishment, Legacies, Solutions* (New York: Routledge, 2018).

19. Environmental Protection Agency, "Strategies for Climate Change Adaptation," accessed September 20, 2019, https://www.epa.gov/arc-x/strategies-climate-change-adaptation.

20. Arun Agrawal, *The Role of Local Institutions in Adaptation to Climate Change* (World Bank, 2008); John Nordgren, Missy Stults, and Sara Meerow, "Supporting Local Climate Change Adaptation: Where We Are and Where We Need to Go," *Environmental Science and Policy* 66 (December 1, 2016): 344–52.

21. David Ciplet, J. Timmons Roberts, and Mizan R. Khan, *Power in a Warming World: The New Global Politics of Climate Change and the Remaking of Environmental Inequality* (Cambridge, MA: MIT Press, 2015), 5.

22. World Commission on Dams, *Dams and Development*; Patrick McCully, *Silenced Rivers: The Ecology and Politics of Large Dams*, 2nd ed. (London: Zed Books, 2001); Sanjeev Khagram, *Dams and Development: Transnational Struggles for Water and Power* (Ithaca, NY: Cornell University Press, 2004).

23. David Schlosberg, *Defining Environmental Justice: Theories, Movements, and Nature* (Oxford: Oxford University Press, 2007); David Schlosberg, "Climate Justice and Capabilities: A Framework for Adaptation Policy," *Ethics and International Affairs* 26, no. 4 (2012): 445–61; Julian Agyeman, David Schlosberg, Luke Craven, and Caitlin Matthews, "Trends and Directions in Environmental Justice: From Inequity to Everyday Life, Community, and Just Sustainabilities," *Annual Review of Environment and Resources* 41, no. 1 (2016): 321–40.

24. For an activist's analysis of the declining effectiveness of protest, see Micah White, *The End of Protest: A New Playbook for Revolution* (Toronto: Alfred A. Knopf Canada, 2016). For an account of how governments have used free-market policies and other tactics to quell overt dissent, see Alasdair Roberts, *The End of Protest: How Free-Market Capitalism Learned to Control Dissent* (Ithaca, NY: Cornell University Press, 2013).

25. For a concise version of Gramsci's text on the war of maneuver and the war of position, as well as a clear description of these and other terms used by him, see Antonio Gramsci, *An Antonio Gramsci Reader: Selected Writings, 1916–1935*, ed. David Forgacs (New York: Schocken Books, 1988), 225–30, 420–31. For lengthier selections of Gramsci's work, see Antonio Gramsci, *Selections from the Prison Notebooks of Antonio Gramsci*, ed. Quintin Hoare and Geoffrey Nowell-Smith (New York: International Publishers, 1971).

26. Gramsci, *Selections from the Prison Notebooks of Antonio Gramsci*.

27. Eve Bratman uses a Gramscian framework in her analysis of why activists failed to halt the construction of Belo Monte. Eve Bratman, "Passive Revolution in the Green Economy: Activism and the Belo Monte Dam," *International Environmental Agreements: Politics, Law and Economics* 15, no. 1 (2015): 61–77.

28. Alexis de Tocqueville, *Democracy in America*, ed. Harvey C. Mansfield and Delba Winthrop (Chicago: University of Chicago Press, 2000); John Stuart Mill, *Considerations on Representative Government* (New York: Harper and Brothers, 1862).

29. Gaventa, "Triumph, Deficit or Contestation?," 3.

30. For a discussion on how and why participatory programs in Latin America emerged and expanded, see Leonardo Avritzer, *Democracy and the Public Space in Latin America* (Princeton, NJ: Princeton University Press, 2002).

31. For more on the defining features of empowered participatory governance, see Archon Fung and Erik Olin Wright, *Deepening Democracy: Institutional Innovations in Empowered Participatory Governance* (London: Verso, 2003).

32. The purposes and effects of participatory democracy have been debated in the scholarly literature since at least the turn of this century. The quoted phrase comes from Caroline W. Lee, Michael McQuarrie, and Edward T. Walker, eds., *Democratizing Inequalities: Dilemmas of the New Public Participation* (New York: New York University Press, 2015). For other examples, see Bill Cooke and Uma Kothari, eds., *Participation: The New Tyranny?* (London: Zed Books, 2001); Andrea Cornwall and Vera Schattan P. Coelho, eds., *Spaces for Change? The Politics of Citizen Participation in New Democratic Arenas* (New York: Zed Books, 2007); Gianpaolo Baiocchi and Ernesto Ganuza, *Popular Democracy: The Paradox of Participation* (Stanford, CA: Stanford University Press, 2016).

33. While much of the scholarship on social movements avoids addressing the role of law and legal mobilization, there is a growing and vibrant body of literature on the use of law by social movements. For some of the influential examples of research on legal mobilization, see Michael W. McCann, *Rights at Work: Pay Equity Reform and the Politics of Legal Mobilization* (Chicago: University of Chicago Press, 1994); Stuart A. Scheingold, *The Politics of Rights: Lawyers, Public Policy, and Political Change*, 2nd ed. (Ann Arbor: University of Michigan Press, 2004); Boaventura de Sousa Santos and César A. Rodríguez-Garavito, eds., *Law and Globalization from Below: Towards a Cosmopolitan Legality* (Cambridge: Cambridge University Press, 2005); Michael W. McCann, ed., *Law and Social Movements*, 2nd ed. (New York: Routledge, 2017).

34. Accounts of state-society relations in democratic and post-neoliberal Latin America have shown how surprising coalitions have formed, how new conflicts have emerged, and why some participatory institutions have taken hold nationally while others have not. Lindsay Mayka, *Building Participatory Institutions in Latin America: Reform Coalitions and Institutional Change* (Cambridge: Cambridge University Press, 2019); Jessica A. J. Rich, *State-Sponsored Activism: Bureaucrats and Social Movements in Democratic Brazil* (Cambridge: Cambridge University Press, 2019); Riofrancos, *Resource Radicals*.

35. Mische, *Partisan Publics*, 50.

36. Baiocchi, Heller, and Silva, *Bootstrapping Democracy*.

1. DAMS AND DEVELOPMENT

1. The World Commission on Dams was a global body of representatives from government, the private sector, and civil society, funded by the World Bank and the World Conservation Union, to assess "the development effectiveness of large dams" and create standards for the planning, construction, monitoring, and decommissioning of dams. World Commission on Dams, *Dams and Development: A New Framework for Decision-Making* (London: Earthscan, 2000), 5.

2. Quoted in David P. Billington and Donald C. Jackson, *Big Dams of the New Deal Era: A Confluence of Engineering and Politics* (Norman: University of Oklahoma Press, 2017), 146.

3. Sukhan Jackson and Adrian Sleigh, "Resettlement for China's Three Gorges Dam: Socio-Economic Impact and Institutional Tensions," *Communist and Post-Communist Studies* 33, no. 2 (June 1, 2000): 223–41.

4. Three Gorges is the largest dam as measured by installed generating capacity, but the Itaipú Dam, on the border of Brazil and Paraguay, sometimes generates more energy on an annual basis than Three Gorges. "Three Gorges Dam: World's Biggest Hydroelectric Facility," U.S. Geological Survey, 2016, https://water.usgs.gov/edu/hybiggest.html.

5. Quoted in Andrew Wyatt, "Building the Temples of Postmodern India: Economic Constructions of National Identity," *Contemporary South Asia* 14, no. 4 (December 2005): 465–80; 465.
6. "India-WRIS Water Resource Project Sub Info System," accessed November 6, 2018, http://india-wris.nrsc.gov.in/wrpapp.html?show=JI00410.
7. Billington and Jackson, *Big Dams of the New Deal Era*; Doug Struck, "Setting Rivers Free: As Dams Are Torn Down, Nature Is Quickly Recovering," *Christian Science Monitor*, August 3, 2014, https://www.csmonitor.com/Environment/2014/0803/Setting-rivers-free-As-dams-are-torn-down-nature-is-quickly-recovering.
8. For more statistics on the number and types of dams in the world, see "International Commission on Large Dams," accessed November 9, 2018, https://www.icold-cigb.org/; Tracy Lane, "World Energy Resources: Charting the Upsurge in Hydropower Development," (London: World Energy Council, 2015.) https://www.worldenergy.org/publications/entry/charting-the-upsurge-in-hydropower-development-2015
9. For a longer discussion of general ecological impacts of dams, see Patrick McCully, *Silenced Rivers: The Ecology and Politics of Large Dams*, 2nd ed. (London: Zed Books, 2001). For more on the debate over the amount of carbon emitted from dams, see McCully, *Silenced Rivers*; Marco Aurelio dos Santos, Luiz Pinguelli Rosa, Bohdan Sikar, Elizabeth Sikar, and Ednaldo Oliveira dos Santos, "Gross Greenhouse Gas Fluxes from Hydro-Power Reservoir Compared to Thermo-Power Plants," *Energy Policy* 34 (2006): 481–88; Jim Giles, "Methane Quashes Green Credentials of Hydropower," *Nature* 444 (November 29, 2006): 524–25; Nathan Barros, Jonathan J. Cole, Lars J. Travnik, Yves T. Prairie, David Bastviken, Vera L. M. Huszar, Paul del Giorgio, and Fábio Roland, "Carbon Emission from Hydroelectric Reservoirs Linked to Reservoir Age and Latitude," *Nature Geoscience* 4 (2011): 593–96; Bridget R. Deemer, John A. Harrison, Siyue Li, Jake L. Beaulieu, Tonya DelSontro, Nathan Barros, José F. Bezerra-Neto, Stephen M. Powers, Marco A. dos Santos, and J. Arie Vonk, "Greenhouse Gas Emissions from Reservoir Water Surfaces: A New Global Synthesis," *BioScience* 66, no. 11 (November 1, 2016): 949–64.
10. The numbers of people displaced by dam construction come from the World Commission on Dams, *Dams and Development*; "International Commission on Large Dams."
11. Sanjeev Khagram, *Dams and Development: Transnational Struggles for Water and Power* (Ithaca, NY: Cornell University Press, 2004); McCully, *Silenced Rivers*; World Commission on Dams, *Dams and Development*.
12. For more information on the conflict over the dams in the Narmada Valley in India, see Amita Baviskar, *In the Belly of the River: Tribal Conflicts over Development in the Narmada Valley* (Delhi: Oxford University Press, 2004); "Women for Forests and Fossil Fuel/Mining/Mega-Dam Resistance | WECAN," accessed November 9, 2018, https://wecaninternational.org/pages/forests-fossil-fuel-resistance. Other in-depth accounts of controversies surrounding specific dam projects include Gustavo Lins Ribeiro, *Transnational Capitalism and Hydropolitics in Argentina: They Yacyretá High Dam* (Gainesville: University Press of Florida, 1994); Allen F. Isaacman and Barbara. S. Isaacman, *Dams, Displacement, and the Delusion of Development: Cahora Bassa and Its Legacies in Mozambique, 1965–2007* (Athens: Ohio University Press, 2013); Jacob Blanc, *Before the Flood: The Itaipu Dam and the Visibility of Rural Brazil* (Durham, NC: Duke University Press, 2019); Ed Atkins, *Contesting Hydropower in the Brazilian Amazon* (New York: Routledge, 2021).
13. For an overview of the controversies surrounding large dams and development, see Khagram, *Dams and Development*. For more discussion on how delays to dam construction can improve outcomes for dam-affected populations and the environment, see Kathryn Hochstetler, "The Politics of Environmental Licensing: Energy Projects of the Past and Future in Brazil," *Studies in Comparative International Development* 46 (October 5, 2011): 349–71.

14. By way of comparison, while Brazil's hydropower generates 80 percent of the country's electricity, hydroelectric dams worldwide generate about 16.4 percent of the world's total electricity. World Energy Resources, "Charting the Upsurge in Hydropower Development," 2015, https://doi.org/10.1002/jbm.b.33207; Empresa de Pesquisa Energética, "Balanço Energético Nacional 2013—Ano Base 2012: Relatório Síntese" (Rio de Janeiro, 2013), http://www.epe.gov.br/pt/publicacoes-dados-abertos/publicacoes/Balanco-Energetico-Nacional-2013. The International Commission on Large Dams (ICOLD) defines a large dam as one that is at least fifteen meters high or has a reservoir that is more than three million cubic meters, if the dam is between five and fifteen meters high. "International Commission on Large Dams."

15. Scholars first used the concept of the developmental state to describe the surprising economic successes of East Asian countries. See Chalmers A. Johnson, *MITI and the Japanese Miracle: The Growth of Industrial Policy, 1925–1975* (Stanford, CA: Stanford University Press, 1982); Alice H Amsden, *Asia's Next Giant: South Korea and Late Industrialization* (New York: Oxford University Press, 1989); Robert Wade, *Governing the Market: Economic Theory and the Role of Government in East Asian Industrialization* (Princeton, NJ: Princeton University Press, 1990); Atul Kohli, *State-Directed Development: Political Power and Industrialization in the Global Periphery* (Cambridge: Cambridge University Press, 2004); Adrian Leftwich, "Bringing Politics Back In: Towards a Model of the Developmental State," *Journal of Development Studies* 31, no. 3 (February 1995): 400–427. For more on how scholars have used the ideas of developmentalism in the Brazilian context, see Kurt Weyland, "From Leviathan to Gulliver? The Decline of the Developmental State in Brazil," *Governance* 11, no. 1 (January 1, 1998): 51–75; Ben Ross Schneider, "The Desarrollista State in Brazil and Mexico," in *The Developmental State*, ed. Meredith Woo-Cumings (Ithaca, NY: Cornell University Press, 1999), 276–305; Kathryn Hochstetler and Ricardo J. Tranjan, "Environment and Consultation in the Brazilian Democratic Developmental State," *Comparative Politics* 48, no. 4 (July 1, 2016): 497–516.

16. McCully, *Silenced Rivers*; Khagram, *Dams and Development*.

17. For more on Itaipú's construction, its impacts, and the resistance to it, see Christine Folch, *Hydropolitics: The Itaipu Dam, Sovereignty, and the Engineering of Modern South America* (Princeton, NJ: Princeton University Press, 2019); Blanc, *Before the Flood*.

18. For more on the extractive development policy and its impacts on local economies, see Stephen G. Bunker, *Underdeveloping the Amazon: Extraction, Unequal Exchange, and the Failure of the Modern State* (Urbana: University of Illinois Press, 1985).

19. For more details on the aluminum industry's use of Tucuruí's power and how the Brazilian government provided expansive subsidies for aluminum smelting firms, see Philip M. Fearnside, "Social Impacts of Brazil's Tucuruí Dam," *Environmental Management* 24, no. 4 (1999): 483–95; Lúcio Flávio Pinto, "De Tucuruí a Belo Monte: A História Avança Mesmo?," *Boletim do Museu Paraense Emílio Goeldi. Ciências Humanas* 7, no. 3 (2012): 777–82.

20. For more details on Turucuí's social and environmental impacts, including the debates over the precise number of displaced people, see Fearnside, "Social Impacts of Brazil's Tucuruí Dam"; Philip M. Fearnside, "Environmental Impacts of Brazil's Tucuruí Dam: Unlearned Lessons for Hydroelectric Development in Amazonia," *Environmental Management* 27, no. 3 (2001): 377–96; Pinto, "De Tucuruí a Belo Monte." For more on the specific and varying estimates of Tucuruí's emissions, see Philip M. Fearnside, "Greenhouse Gas Emissions from a Hydroelectric Reservoir (Brazil's Tucuruí Dam) and the Energy Policy Implications," *Water, Air, and Soil Pollution* 133 (2002): 69–96; Marcelo Pedroso Curtarelli, Igor Ogashawara, Carlos Alberto Sampaio de Araújo, João Antônio Lorenzzetti, Joaquim Antônio Dionísio Leão, Enner Alcântara, José Luiz Stech, "Carbon Dioxide Emissions from Tucuruí Reservoir (Amazon Biome): New Findings Based on Three-Dimensional Ecological Model Simulations," *Science

of the Total Environment 551–52 (2016): 676–94. For the images and reports of flooded vegetation, see Leonardo Coutinho, "Bilhões Embaixo d'agua," *Veja*, June 2004; Paula Sampaio, "O Lago do Esquecimento," 2012, http://paulasampaio.com.br/projetos/lago-do-esquecimento/.

21. Henri Acselrad, "Planejamento Autoritário e Desordem Socioambiental na Amazônia: Crônica do Deslocamento de Populações Em Tucuruí," *Revista de Administração Pública* 25, no. 4 (1991): 68.

22. For more on the benefits, costs, and impacts of Tucuruí, the ways and reasons decisions were made about the dam, changes from the project's plans, and the "development effectiveness" of Tucuruí, see E. L. La Rovere and F. E. Mendes, "Tucuruí Hydropower Complex, Brazil, A WCD Case Study," *World Commission on Dams* (Cape Town, 2000). Additionally, for a list of specific cases of injustice and human rights violations attributed to Tucuruí's construction, see the report drafted by the Brazilian Council for the Defense of Human Rights, which traveled to Tucuruí to investigate complaints made against the dam by the nationwide group Movimento de Atingidos por Barragens (MAB—Movement of People Affected by Dams). Conselho de Defesa dos Direitos da Pessoa Humana, "Comissão Especial: 'Atingidos Por Barragens'" (Brasília, 2007). Also see Philip M. Fearnside, "Social Impacts of Brazil's Tucuruí Dam," *Environmental Management* 24, no. 4 (1999): 483–95; Fearnside, "Environmental Impacts of Brazil's Tucuruí Dam; Lúcio Flávio Pinto, "De Tucuruí a Belo Monte."

23. Balbina's installed capacity is just 250 megawatts. Tucuruí's initial capcity was 4,000 megawatts, but after the second stage of construction was complete, it now has 25 turbines, with a total generative capacity of over 8,000 megawatts.

24. Barbara Cummings, Philip Fearnside, and Stephen Baines document the social and environmental damage of Balbina, provide details on the displacement of Indigenous and other communities, and convincingly show how the project was an economic and technological folly. Barbara J. Cummings, *Dam the Rivers, Damn the People: Development and Resistence in Amazonian Brazil*, 2nd ed. (London: Earthscan Publications, 2009); Philip M. Fearnside, "Brazil's Balbina Dam: Environment versus the Legacy of the Pharaohs in Amazonia," *Environmental Management* 13, no. 4 (1989): 401–23; Stephen Grant Baines, "A Usina Hidrelétrica de Balbina e o Deslocamento Compulsório dos Waimiri-Atroari," *Série Antropologia* 166 (1994). For a detailed description of a scientific study on carbon emissions released from the Balbina dam, see Alexandre Kemenes, Bruce R. Forsberg, and John M. Melack, "CO_2 Emissions from a Tropical Hydroelectric Reservoir (Balbina, Brazil)," *Journal of Geophysical Research* 116, no. 3 (2011): 1–11.

25. For a detailed account of military rule and the slow return to democracy in the 1970s and 1980s, see Thomas E. Skidmore, *The Politics of Military Rule in Brazil, 1964–85* (New York: Oxford University Press, 1988).

26. For more on the history of MAB, see Franklin Daniel Rothman and Pamela E. Oliver, "From Local to Global: The Anti-Dam Movement in Southern Brazil, 1979–1992," *Mobilization: An International Quarterly* 4, no. 1 (1999): 41–57; Sabrina McCormick, "The Brazilian Anti-Dam Movement: Knowledge Contestation as Communicative Action," *Organization and Environment* 19, no. 3 (September 1, 2006): 321–46; "A Criação das Comissões Regionais de Atingidos," Movimento dos Atingidos por Barragens, 2011, http://www.mabnacional.org.br/content/2-cria-das-comiss-es-regionais-atingidos.

27. Philip M. Fearnside, "Dams in the Amazon: Belo Monte and Brazil's Hydroelectric Development of the Xingu River Basin," *Environmental Management* 38, no. 1 (July 2006): 16–27; Zachary Hurwitz, Brent Millikan, Telma Monteiro, and Roland Widmer *Mega-Projeto, Mega-Riscos: Análise de Riscos Para Investidores No Complexo Hidrelétrico Belo Monte*, (São Paulo: Amigos da Terra—Amazônia Brasileira and International Rivers, 2011); Carlos Alberto de Moya, Helio Costa de Barros Franco, and Paulo Fernando Vieira Souto Rezende, "AHE Belo

Monte-Evolução dos Estudos," in *XXVII Seminário Nacional de Grandes Barragens* (Belém: Comitê Brasileiro de Barragens, 2007).

28. Moya, Franco, and Rezende, "AHE Belo Monte-Evolução dos Estudos."

29. For more details on the changes to the plans to dam the Xingu River, see Fearnside, "Dams in the Amazon."

30. For more technical details on the Belo Monte dam, including power generation figures and the engineering design, see the following document published by Norte Energia officials: O. M. Bandeira and J. B. de Menezes, "Current Progress at the Belo Monte Hydro Project, Brazil," *Hydropower and Dams*, no. 6 (2018): 29–39. International Rivers, an organization that works to stop the construction of large dams around the world, and other anti-dam groups have focused some of their arguments against Belo Monte on the dam's variable energy generation and the relatively small amounts of electricity generated for upward of six months every year. See International Rivers, "Belo Monte: Massive Dam Project Strikes at the Heart of the Amazon" (Berkeley, CA: 2012).

31. The engineer is quoted in Philp Fearnside's description of the emotional and logical reasons that proponents have used to advocate for the construction of a hydroelectric facility on the Xingu River. Fearnside, "Dams in the Amazon," 19.

32. For more on the details of PAC, see the government's website that details the program. "Sobre o PAC," accessed June 27, 2019, http://www.pac.gov.br/sobre-o-pac.

33. For more on the energy shortage in 2001, electricity rationing, and how this event became important in the rhetoric around hydroelectric development and the construction of Belo Monte, see Ildo Luís Sauer, José Paulo Vieira, and Carlos Augusto Ramos Kirchner, *O Racionamento de Energia Elétrica Decretado Em 2001: Um Estudo sobre as Causas e as Responsabilidades*, (São Paulo: Instituto de Energia e Ambiente/Universidade de São Paulo, 2001); https://repositorio.usp.br/item/001224797 Georgia O. Carvalho, "Environmental Resistance and the Politics of Energy Development in the Brazilian Amazon," *Journal of Environment and Development* 15, no. 3 (2006): 245–68; Fearnside, "Dams in the Amazon"; Eve Z. Bratman, "Contradictions of Green Development: Human Rights and Environmental Norms in Light of Belo Monte Dam Activism," *Journal of Latin American Studies* 46, no. 2 (May 2014): 261–89.

34. In order to describe the Brazilian state's approach to development, I use the conceptual term "democratic developmentalism" that Gordon White and Mark Robinson advanced in the 1990s. Scholars have also used other descriptions to explain the reemergence of a state-led, more socially just and inclusive model of development in Latin America and Brazil, such as the "new developmental welfare state," "new state activism," and "neostructuralism." All of these concepts capture the reality that Brazil's approach under the Workers' Party was neither completely new nor was it the same as in the past. While the terminology is not a critical issue, a few other scholars and I have suggested that it is helpful to describe Brazil as a democratic developmental state because it encourages examination along two fronts: the state's ability to implement deeper democratic practices and its simultaneous capacity to encourage and guide the infrastructure development that it desires. Gordon White, "Towards a Democratic Developmental State," *IDS Bulletin* 26, no. 2 (1995): 27–36; Mark Robinson and Gordon White, eds., *The Democratic Developmental State: Politics and Institutional Design*, Oxford Studies in Democratization. (New York: Oxford University Press, 1998); Sônia M. Draibe and Manuel Riesco, "Latin America: A New Developmental Welfare State Model in the Making?," in *Latin America: A New Developmental Welfare State Model in the Making?*, ed. Manuel Riesco (Basingstoke: Palgrave, 2007), 21–113; David M. Trubek, Diogo R. Coutinho, and Mario G. Schapiro, "New State Activism in Brazil and the Challenge for Law," in *Law and the New Developmental State: The Brazilian Experience in Latin American Context*, ed. David M. Trubek, Helena Alviar Garcia, Diogo R. Coutinho, and Alvaro Santos (Cambridge: Cambridge University

Press, 2013); Fernando Ignacio Leiva, *Latin American Neostructuralism: The Contradictions of Post-Neoliberal Development* (Minneapolis: University of Minnesota Press, 2008); Peter Taylor Klein, "Engaging the Brazilian State: The Belo Monte Dam and the Struggle for Political Voice," *Journal of Peasant Studies* 42, no. 6 (March 20, 2015): 1–20; Grant Burrier, "The Developmental State, Civil Society, and Hydroelectric Politics in Brazil," *Journal of Environment and Development* 25, no. 3 (2016): 1–27; Hochstetler and Tranjan, "Environment and Consultation in the Brazilian Democratic Developmental State."

35. An extensive literature on Bolsa Família documents the largely positive impacts of the cash-transfer program, while also showing how the effects have been differentiated by gender and class, how the program could lead to unforeseen negative consequences in the long term, and how Bolsa Família could open pathways to new forms of clientelism. See Anthony Hall, "Brazil's Bolsa Família: A Double-Edged Sword?," *Development and Change* 39, no. 5 (2008): 799–822; Edmund Amann and Werner Baer, "The Macroeconomic Record of the Lula Administration, the Roots of Brazil's Inequality, and Attempts to Overcome Them," in *Brazil under Lula*, ed. Joseph L. Love and Werner Baer (New York: Palgrave Macmillan, 2009), 27–43; Fábio Veras Soares, Rafael Perez Ribas, and Rafael Guerreiro Osório, "Evaluating the Impact of Brazil's Bolsa Família: Cash Transfer Programs in Comparative Perspective," *Latin American Research Review* 45, no. 2 (2010): 173–90; Alan de Brauw, Daniel O. Gilligan, John Hoddinott, and Shalini Roy, "The Impact of Bolsa Família on Schooling," *World Development* 70 (2015): 303–16; Gregory Duff Morton, "Managing Transience: Bolsa Família and Its Subjects in an MST Landless Settlement," *Journal of Peasant Studies* 42, no. 6 (November 2, 2015): 1283–1305; Jonathan Tepperman, "Brazil's Antipoverty Breakthrough: The Surprising Success of Bolsa Família," *Foreign Affairs* 95, no. 1 (2016): 34–47.

36. Gianpaolo Baiocchi, *Militants and Citizens: The Politics of Participatory Democracy in Porto Alegre* (Stanford, CA: Stanford University Press, 2005); Gianpaolo Baiocchi, Patrick Heller, and Marcelo Silva, *Bootstrapping Democracy: Transforming Local Governance and Civil Society in Brazil* (Stanford, CA: Stanford University Press, 2011).

37. For detailed analysis of Brazil's ability to usher in nationwide participatory institutions, see Lindsay Mayka, *Building Participatory Institutions in Latin America: Reform Coalitions and Institutional Change* (Cambridge: Cambridge University Press, 2019). For more on the spread, use, and effectiveness of management councils throughout Brazil, see Vera Schattan P. Coelho, "Brazilian Health Councils: Including the Excluded?," in *Spaces for Change: The Politics of Citizen Participation in New Democratic Arenas*, ed. Andrea Cornwall and Vera Schattan P Coelho (New York: Zed Books, 2007), 33–54; Leonardo Avritzer, *Participatory Institutions in Democratic Brazil* (Baltimore, MD: Johns Hopkins University Press, 2009); Christopher L. Gibson, *Movement-Driven Development: The Politics of Health and Democracy in Brazil* (Stanford, CA: Stanford University Press, 2019).

38. Some scholars suggest that local participatory initiatives were successful because they were based on the logics of "sharing power" and "empowerment," whereas the logics of the federally instituted initiatives became "listening" and "dialogue." See Gianpaolo Baiocchi, Einar Braathen, and Ana Claudia Teixeira, "Transformation Institutionalized? Making Sense of Participatory Democracy in the Lula Era," in *Democratization in the Global South: The Importance of Transformative Politics*, ed. Kristian Stokke and Olle Törnquist (New York: Palgrave Macmillan, 2013), 217–39. Others point to the lack of presidential advisors involved in radical democratic participation, the strength of opposition parties, and President Lula's inability to learn from the experiences of participatory budgeting as reasons for the failures of participatory institutions at the national level. See Benjamin Goldfrank, "The Left and Participatory Democracy: Brazil, Uruguay, and Venezuela," in *The Resurgence of the Latin American*

Left, ed. Steven Levitsky and Kenneth M. Roberts (Baltimore, MD: Johns Hopkins University Press, 2011), 162–83.

39. Brasil and Presidência da República, "Plano Amazônia Sustentável: Diretrizes Para o Desenvolvimento Sustentável da Amazônia Brasileira" (Brasília, 2008).

40. Philip M. Fearnside, "Brazil's Cuiabá-Santarém (BR-163) Highway: The Environmental Cost of Paving a Soybean Corridor through the Amazon," *Environmental Management* 39 (2007): 601–14; Jeremy M. Campbell, *Conjuring Property: Speculation and Environmental Futures in the Brazilian Amazon* (Seattle: University of Washington Press, 2015).

41. Presidência da República—Casa Civil, "Decreto 7.340, de 21.10.2010" (2010).

42. Heather Randell, "The Short-Term Impacts of Development-Induced Displacement on Wealth and Subjective Well-Being in the Brazilian Amazon," *World Development* 87 (November 1, 2016): 385–400.

43. According to research carried out by Globo, South America's largest commercial television network, the homicide rate in Altamira in 2015 was 124.6 homicides per 100,000 inhabitants. By comparison, the rates in Rio de Janeiro and São Paulo were 23.4 and 13.5, respectively, in the same year. See Danielle Nogueira, "Altamira: A Vida na Cidade Mais Violenta do Brasil - Jornal O Globo," Globo, 2017, https://oglobo.globo.com/brasil/altamira-vida-na-cidade-mais-violenta-do-brasil-22183157. See also "Ipea—Atlas da Violencia—Mapa," Instituto de Pesquisa Econômica Aplicada, accessed January 3, 2019, http://www.ipea.gov.br/atlasviolencia/dados-series/20.

44. According to the Brazilian census of 2010, the population of Altamira was 99,075, with 84,092 people living in the urban area. It is nearly impossible to obtain an accurate population count for the years during dam construction, due to the constant flux of people in and out of the area, but conservative estimates suggest that the urban population grew by at least 50 percent from 2010 to 2015. Research on the physical expansion of the city supports these claims, showing that the city footprint grew by nearly 79 percent from 2011 to 2015. For official population counts, see "Brasil Em Síntese," Instituto Brasileiro de Geografia e Estatística, accessed January 4, 2019, https://cidades.ibge.gov.br/brasil/pa/altamira/pesquisa/23/27652. For population growth estimates and the physical expansion of the urban areas, see José Quieroz de Miranda Neto and José Antônia Herrera, "Expansão Urbana Recente Em Altamira (PA): Novas Tendências de Screscimento a Partir da Instalação da UHE Belo Monte," *Ateliê Geográfico* 11, no. 3 (2017): 34–52; Davieliton Mesquita Pinho, Gustavo Carvalho Spanner, Genilson Santana Cornélio, José Antônio Herrera, and Alan Nune Araújo, "Evolução da Mancha Urbana do Município de Altamira, Pará," in *Anais do II Congresso Amazônico de Meio Ambiente e Energias Renováveis: Engenharia, Meio Ambiente e Desenvolvimento Energético*, UFRA Campus Belém, Pará, Brazil, September 12–16, 2016, 1078–1086. https://www.even3.com.br/anais/camaer2016/ https://www.even3.com.br/anais/camaer2016/31492-evolucao-da-mancha-urbana-do-municipio-de-altamira-para/

45. For more on *barrageiro* culture, see Fearnside, "Brazil's Balbina Dam." For an example of prostitution in the contexts of large projects in Brazil, see W.R.M. Araújo, "Sociabilidades, Trabalho Sexual, e Autonomia No Contexto das Grandes Obras na Amazônia Brasileira," *Revista Relações Sociais* 1, no. 3 (2018): 523–36. For more on the trafficking related to the Belo Monte project, see Zachary Hurwitz, "Sex Trafficking Ringmaster Busted on Belo Monte," *International Rivers*, September 3, 2013.

46. Research conducted in Altamira on the impacts of the population growth associated with Belo Monte's construction confirm my ethnographic observations. Douglas Pereira de Souza, Wanhinna Regina Soares da Silva, Gilberto Carlos Cervinski, Bruno Dias dos Santos, Francisco de Assis Comarú, and Federico Bernardino Morante Trigoso, "Desenvolvimento Urbano e

Saúde Pública: Impactos da Construção da UHE de Belo Monte," *Desenvolvimento e Meio Ambiente* 46 (2018): 2176–9109; Neto and Herrera, "Expansão Urbana Recente Em Altamira (PA)."

2. BOOMS, BUSTS, AND COLLECTIVE MOBILIZATION ALONG THE TRANSAMAZON

1. The theory of unequal exchange, as developed in Stephen G. Bunker, *Underdeveloping the Amazon: Extraction, Unequal Exchange, and the Failure of the Modern State* (Urbana: University of Illinois Press, 1985), describes how the extraction of resources in Amazonia has benefited distant economies and dominant classes, while stifling social and economic development in the region where extraction takes place. A rich literature on "ecologically unequal exchange" has developed that draws on Bunker's theory, world-systems approach, and Marxist ideas of the metabolic rift. This literature shows how the structure of the global political economy, which is largely based on unequal material and ecological exchanges, has created ongoing social and environmental inequities. Thomas K. Rudel, J. Timmons Roberts, and JoAnn Carmin, "The Political Economy of the Environment," *Annual Review of Sociology* 37 (2011): 221–38; Andrew K. Jorgenson, "Environment, Development, and Ecologically Unequal Exchange," *Sustainability (Switzerland)* 8, no. 3 (2016); Brett Clark and John Bellamy Foster, "Ecological Imperialism and the Global Metabolic Rift," *International Journal of Comparative Sociology* 50, no. 3–4 (2009): 311–34.
2. William M. Denevan, "The Aboriginal Population of Amazonia," in *The Native Population of the Americas in 1492*, ed. William M. Denevan (Madison: University of Wisconsin Press, 1992), 205–34.
3. For more on the history of rubber in the Amazon, see Warren Dean, *Brazil and the Struggle for Rubber: A Study in Environmental History* (Cambridge: Cambridge University Press, 1987). For more on the settlement of the Xingu region and the founding of Altamira, see Marianne Schmink and Charles H. Wood, *Contested Frontiers in Amazonia* (New York: Columbia University Press, 1992); Emilio F. Moran, *Developing the Amazon* (Bloomington: Indiana University Press, 1981); Antônio Ubirajara Bogea Umbuzeiro and Ubirajara Marques Umbuzeiro, *Altamira e Sua História*, 4th ed. (Belém: Ponto Press, 2012).
4. While the market for Amazonian rubber dwindled by the 1920s, the Ford Motor Company did not abandon their rubber plantations along the Tapajós River until 1945. Dean, *Brazil and the Struggle for Rubber*; Greg Grandin, *Fordlandia: The Rise and Fall of Henry Ford's Forgotten Jungle City* (New York: Metropolitan Books, 2009).
5. "Censo Demográfico: 1960" (Rio de Janeiro, 1968).
6. For more information on the monetization of the region and the lack of federal investment, see Schmink and Wood, *Contested Frontiers in Amazonia*; Moran, *Developing the Amazon*.
7. Moran, *Developing the Amazon*; Umbuzeiro and Umbuzeiro, *Altamira e Sua História*.
8. Charles H. Wood and Marianne Schmink, "The Military and the Environment in the Brazilian Amazon," *Journal of Political and Military Sociology* 21, no. 1 (Summer 1993): 81–105; Thomas E. Skidmore, *The Politics of Military Rule in Brazil, 1964–85* (New York: Oxford University Press, 1988).
9. Wood and Schmink, "The Military and the Environment in the Brazilian Amazon," 85. For more on the military dictatorship's expansion into Amazonia, also see Dennis J. Mahar, *Frontier Development Policy in Brazil: A Study of Amazonia* (New York: Praeger Publishers, 1979).
10. The Transamazon highway was never completed and ends approximately four hundred miles from the border with Peru and Colombia. For more on the Transamazon highway and colonization project, see Mahar, *Frontier Development Policy in Brazil*; Schmink and Wood,

Contested Frontiers in Amazonia; John O. Browder and Brian J. Godfrey, *Rainforest Cities: Urbanization, Development, and Globalization of the Brazilian Amazon* (New York: Columbia University Press, 1997).

11. In addition to rapid urban growth in the early 1970s, the population in the rural regions along the Transamazon also increased significantly. The population of the municipality of Altamira rose from just under 8,000 in 1950 to just under 12,000 in 1960. By 1970, the population was still a modest 15,345, but during the subsequent decade, the number of people tripled to over 46,000 and continued climbing until it reached 72,408 in the census of 1991. Population increased at a more moderate pace for the next twenty years, reaching 77,439 inhabitants in 2000 and 99,075 in 2010. All census data available at "IBGE Biblioteca," Instituto Brasileiro de Geografia e Estatística, 2019, https://biblioteca.ibge.gov.br. For details on changes to Altamira's infrastructure in the 1970s, see Moran, *Developing the Amazon*.

12. For an extended discussion on the multifaceted reasons for the failure of the Transamazon colonization program, as well as the government's inattention to local needs, see Bunker, *Underdeveloping the Amazon*; Moran, *Developing the Amazon*; Schmink and Wood, *Contested Frontiers in Amazonia*.

13. Leonardo Boff and Clodovis Boff are brothers who, after publishing their book, became divided over their beliefs. Leonardo continued to ardently defend liberation theology, while Clodovis became more critical of the perspective. Leonardo Boff and Clodovis Boff, *Introducing Liberation Theology* (Tunbridge Wells: Burns and Oates/Search Press, 1987), 3.

14. For extended discussions on the roots of liberation theology in Brazil and the political and ideological connections and debates between Marxism and liberation theology, see Gustavo Gutiérrez, *A Theology of Liberation: History, Politics, and Salvation*, ed. Sister Caridad Inda and John Eagleson (Maryknoll, NY: Orbis Books, 1973); Phillip Berryman, *Liberation Theology: Essential Facts about the Revolutionary Movement in Latin America—and Beyond* (Philadelphia: Temple University Press, 1987); Michael Löwy and Claudia Pompan, "Marxism and Christianity in Latin America," *Latin American Perspectives* 20, no. 4 (1993): 28–42; Sergio Lessa and Laurence Hallewell, "The Situation of Marxism in Brazil," *Latin American Perspectives* 25, no. 1 (1998): 94–108; Boff and Boff, *Introducing Liberation Theology*. For a broader discussion on the use of Christian Base Communities in Brazil, see John Burdick, "Rethinking the Study of Social Movements: The Case of Christian Base Communities in Urban Brazil," in *The Making of Social Movements in Latin America: Identity, Strategy, and Democracy*, ed. Arturo Escobar and Sonia E. Alvarez (Boulder, CO: Westview Press, 1992), 171–84.

15. Ana Carolina Alfinito Vieira and Sigrid Quack, "Trajectories of Transnational Mobilization for Indigenous Rights in Brazil," *Revista de Administracao de Empresas* 56, no. 4 (2016): 380–94; "Conselho Indigenista Missionário," accessed June 10, 2021, https://cimi.org.br/.

16. A territorial prelature of the Roman Catholic Church is similar to a diocese. It is a territory over which a prelate—Bishop Kräulter, in the case of the Prelazia do Xingu—has jurisdiction. The Prelazia do Xingu—which was promoted and renamed Diocese do Xingu-Altamira in November of 2019—was composed of the priests in sixteen municipalities of the Xingu region. For details on the Prelazia do Xingu, see a "Prelazia do Xingu," Conferência Nacional dos Bispos do Brasil Regional Norte II, accessed July 18, 2019, http://cnbbn2.com.br/prelazia-do-xingu/. For more details on territorial prelatures in general and what the Catholic Church allows and does not allow prelatures to do, see William Fanning, "Praelatus Nullius," in *The Catholic Encyclopedia* (Robert Appleton Company, 1911).

17. "FVPP: A História do Movimento Pelo Desenvolvimento da Transamazônica e Xingu" (Brasília, 2006).

18. Bishop Kräutler discussed the early protests during interviews I conducted with him, but the quote included here is in Paula Lacerda's work on sexual violence against boys in Altamira

during the early 1990s. Paula Mendes Lacerda, *Meninos de Altamira: Violência, "Luta" Política e Administração Pública* (Rio de Janeiro: Garamond, 2015), 113.

19. "Universidade Federal do Para, Campus de Altamira, Histórico do Campus," 2019, https://altamira.ufpa.br/index.php/historico.

20. For detailed information about Indigenous communities around Brazil, including current practices, historical information, and their political struggles, see Beto Ricardo and Fany Ricardo, eds., *Povos Indígenas no Brasil 2006/2010* (São Paulo: Instituto Socioambiental, 2011); Instituto Socioambiental, "Povos Indígenas no Brasil," accessed June 10, 2021, https://pib.socio ambiental.org/pt/Página_principal. For information about the role of Indigenous people in the Brazilian anti-dam movement, see Sabrina McCormick, "The Brazilian Anti-Dam Movement: Knowledge Contestation as Communicative Action," *Organization and Environment* 19, no. 3 (September 1, 2006): 321–46.

21. For the full 2010 Plan, see Eletrobrás, *Plano 2010: Relatório Geral, Plano National de Energia Elétrica 1987/2010* (Rio de Janeiro: Centrais Elétricas Brasileiras S.A., 1987). See also Philip M. Fearnside, "Dams in the Amazon: Belo Monte and Brazil's Hydroelectric Development of the Xingu River Basin," *Environmental Management* 38, no. 1 (2006): 16–27.

22. Dom Erwin Kräutler, "Mensagem de Abertura," in *Tenotã-Mõ: Alertas sobre as Conseqüências dos Projetos Hidrelétricos no Rio Xingu*, ed. A. Oswaldo Sevá Filho (International Rivers Network, 2005), 11.

23. Quoted in Patrick McCully, *Silenced Rivers: The Ecology and Politics of Large Dams*, 2nd ed. (London: Zed Books, 2001), 293. See also A. Oswaldo Sevá Filho, ed., *Tenotã- Mõ: Alertas sobre as Conseqüências dos Projetos Hidrelétricos no Rio Xingu* (International Rivers Network, 2005).

24. Scholars have discussed the shrinking of the developmental state in the 1980s and 1990s, while also highlighting how previous approaches were not abandoned. Kathryn Hochstetler and Alfred Montero, in particular, argue that the large state apparatus that had grown over the previous five decades was altered, but not completely replaced. See Kathryn Hochstetler and Alfred P. Montero, "The Renewed Developmental State: The National Development Bank and the Brazil Model," *Journal of Development Studies* 49, no. 11 (July 17, 2013): 1–16; Kurt Weyland, "From Leviathan to Gulliver? The Decline of the Developmental State in Brazil," *Governance* 11, no. 1 (January 1, 1998): 51–75; Peter Kingstone, "Critical Issues in Brazil's Energy Sector: The Long (and Uncertain) March to Energy Privatization in Brazil," James A. Baker III Institute for Public Policy of Rice University, March, 2004. https://www.bakerinstitute.org/research/the-long-and-uncertain-march-to-energy-privatization-in-brazil/

25. Scholars have argued that multinational development banks (MDBs), particularly the World Bank, gained significant political influence in the energy sector during the 1980s because they were the only source of financing during the Latin American debt crisis. Instead of funding large projects as in the past, the MDBs encouraged economic liberalization and private-sector participation. This approach meant that state-led firms could no longer acquire loans for large projects such as dams. See Edmar Luiz Fagundes de Almeida and Helder Queiroz Pinto Junior, "Driving Forces of the Brazilian Electricity Industry Reform," *Energy Studies Review* 9, no. 2 (1999): 50–65; Franklin D. Rothman, "A Comparative Study of Dam-Resistance Campaigns and Environmental Policy in Brazil," *Journal of Environment and Development* 10, no. 4 (2001): 317–44.

26. Hochstetler and Montero, "The Renewed Developmental State"; Mahrukh Doctor, "Assessing the Changing Roles of the Brazilian Development Bank," *Bulletin of Latin American Research* 34, no. 2 (2014): 197–213.

27. Scholars disagree on the extent to which the transition to a democratic government in the 1980s changed the approach to development and environmental concerns in the Amazon. For arguments that democracy allowed environmentalists, Indigenous communities, and others

to better organize, see Maria Carmen Lemos and J. Timmons Roberts, "Environmental Policy-Making Networks and the Future of the Amazon," *Philosophical Transactions of the Royal Society B* 363 (2008): 1897–902. Others identify the openings for socioenvironmental struggles, yet also highlight the difficulties, particularly within electoral politics, that environmentalists faced. See Kathryn Hochstetler and Margaret E. Keck, *Greening Brazil: Environmental Activism in State and Society* (Durham, NC: Duke University Press, 2007). Less-positive accounts suggest that few concrete results came about as a result of the legislation that sought to codify social and environmental concerns into law. See, for example, Cynthia S. Simmons, "The Local Articulation of Policy Conflict: Land Use, Environment, and Amerindian Rights in Eastern Amazonia," *Professional Geographer* 54, no. 2 (2002): 241–58. For more on environmental impact assessments in Brazil and new socio-environmental regulations, see Hochstetler and Keck, *Greening Brazil*; Kathryn Hochstetler, "The Politics of Environmental Licensing: Energy Projects of the Past and Future in Brazil," *Studies in Comparative International Development* 46 (October 5, 2011): 349–71; Luis E. Sánchez, "Development of Environmental Impact Assessment in Brazil," *Schwerpunktthema* 27, no. 4–5 (2013): 193–200.

28. In 1987, the United Nations World Commission on Environment and Development released *Our Common Future*, also known as the "Brundtland Report." The report provided one of the first and most widely used definitions of "sustainable development." *Our Common Future* became a framework for addressing global issues at gatherings of the United Nations, including the 1992 United Nations Conference on Environment and Development, also known as the Rio de Janeiro Earth Summit. World Commssion on Environment and Development, ed., *Our Common Future: The World Commission on Environment and Development* (Oxford: Oxford University Press, 1987).

29. Most scholars agree that a lack of funding shelved the original plans to build dams on the Xingu. The disagreement over the reasons why the World Bank withdrew financing and why no other lenders would support the project generally relates to how much weight observers give to the organized resistance versus broader structural factors. Some researchers and activists, such as those who authored *Tenotã-Mõ* and others who wrote *Mega-Projeto, Mega-Riscos* point to the sustained mobilization and protests in 1989. Filho, *Tenotã-Mõ*; Zachary Hurwitz, Brent Millikan, Telma Monteiro, and Roland Widmer, *Mega-Projeto, Mega-Riscos: Análise de Riscos Para Investidores no Complexo Hidrelétrico Belo Monte* (São Paulo: Amigos da Terra—Amazônia Brasileira and International Rivers, 2011). Sabrina McCormick, Sanjeev Khagram, and Georgia Carvalho similarly attribute the stoppage to mobilizations, highlighting how grassroots activists, Indigenous people, environmental groups, nongovernmental organizations, and experts collaborated to stop construction. McCormick, "The Brazilian Anti-Dam Movement"; Sanjeev Khagram, *Dams and Development: Transnational Struggles for Water and Power* (Ithaca, NY: Cornell University Press, 2004); Georgia O. Carvalho, "Environmental Resistance and the Politics of Energy Development in the Brazilian Amazon," *Journal of Environment and Development* 15, no. 3 (2006): 245–68. Others, such as Kathryn Hochstetler, acknowledge the important role that mobilizations and anti-dam coalitions played in stalling the project, but also point to the broader national and international factors that led to the World Bank withdrawing its funding. Hochstetler, "The Politics of Environmental Licensing."

30. Scholars have detailed innovative approaches to sustainable development and how local populations throughout the Amazon mobilized and engaged during and after the transition to democracy. These researchers show that, with new national and international attention to social and environmental issues, local people proactively defended their land, collaborated with outside groups, and developed place-based conservation plans to respond to destructive extractive processes. See Anthony Hall, *Sustaining Amazonia: Grassroots Action for Productive Conservation* (Manchester: Manchester University Press, 1997); Anthony L. Hall, ed., *Amazonia* *at the Crossroads: The Challenge of Sustainable Development* (London: University of London, Institute of Latin American Studies, 2000); Hochstetler and Keck, *Greening Brazil*.

31. Paula Lacerda describes in more detail how mobilization through the Catholic Church inspired women to make particular types of claims in Altamira and along the Transamazon. Paula Mendes Lacerda, "Movimentos Sociais na Amazônia: Articulações Possíveis entre Gênero, Religião e Estado," *Boletim do Museu Paraense Emílio Goeldi. Ciências Humanas* 8, no. 1 (2013): 153–68.

32. Maria Ivonete Coutinho da Silva, "Mulheres Migrantes na Transamazônica: Construção da Ocupação e do Fazer Política," PhD diss., (Universidade Federal do Pará, 2008), 239.

33. Paula Lacerda shows how the "case" and the "cause" of emasculation, which she insightfully differentiates in her account of the issue, remained important over the subsequent two decades. She details how social movements continued to hold street rallies and public remembrances in order to draw attention to both ongoing violence and an unjust policing and judicial system. Lacerda, *Meninos de Altamira*.

34. For three of the first in-depth examinations of the Workers' Party formation and rise to power, see Emir Sader and Ken Silverstein, *Without Fear of Being Happy: Lula, the Workers Party and Brazil* (London: Verso, 1991); Gay W. Seidman, *Manufacturing Militance: Workers' Movements in Brazil and South Africa, 1970–1985* (Berkeley: University of California Press, 1994); Margaret E. Keck, *The Workers' Party and Democratization in Brazil* (New Haven, CT: Yale University Press, 1995).

35. For detailed histories of collective organizing and social movements along the Transamazon through the MPST and FVPP, see Ana Paula Santos Souza, "O Desenvolvimento Socioambiental na Transamazônica: A Trajetória de um Discurso a Muitas Vozes" (Universidade Federal do Pará, 2006); "FVPP: A História do Movimento Pelo Desenvolvimento da Transamazônica e Xingu"; Fabiano Toni, "Institutional Choices on the Brazilian Agricultural Frontier: Strengthening Civil Society or Outsourcing Centralized Natural Resource Management," (paper presented at the *Eleventh Conference of the International Association for the Study of Common Property*, Bali, June 19-23, 2006), https://hdl.handle.net/10535/583.

36. "FVPP: A História do Movimento Pelo Desenvolvimento da Transamazônica e Xingu"; Souza, "O Desenvolvimento Socioambiental na Transamazônica"; Eve Z. Bratman, "Contradictions of Green Development: Human Rights and Environmental Norms in Light of Belo Monte Dam Activism," *Journal of Latin American Studies* 46, no. 2 (May 2014): 261–89.

37. Two of the edited volumes produced as a result of collaborative processes between researchers and activists include *Tenotã-Mõ* and the report issued by the Panel of Specialists Filho, *Tenotã- Mõ*; Sônia Barbosa Magalhães and Francisco del Moral Hernández, eds., *Painel de Especialistas: Análise Crítica do Estudo de Impacto Ambiental do Aproveitamento Hidrelétrico de Belo Monte*, (Belém, 2009).

38. A rich scholarly literature examines the connections between the Workers' Party and social movements. Margaret Keck shows how the Workers' Party began from a militant trade union and then expanded to be a party that appealed to a broad constituency of people who felt excluded and marginalized from Brazil's economic and social structures. Keck, *The Workers' Party and Democratization in Brazil*. Angus Wright and Wendy Wolford, among others, focus specifically on the Landless Workers' Movement (MST—Movimento dos Trabalhadores Rurais Sem Terra), one of the largest social movements in Latin America, and include lengthy discussions on the relationships between the MST and the Workers' Party. Angus Wright and Wendy Wolford, *To Inherit the Earth: The Landless Movement and the Struggle for a New Brazil* (Oakland, CA: Food First Books, 2003). David Samuels shows the internal institutional changes the Workers' Party made in order to win the presidential election and gain power throughout the country, highlighting how the character of the party changed from movement

to mainstream political entity. David Samuels, "From Socialism to Social Democracy: Party Organization and the Transformation of the Workers' Party in Brazil," *Comparative Political Studies* 37, no. 9 (2004): 999–1024.

3. DEMOCRATIC DEVELOPMENTALISM

1. This is my translation of President Lula's speech, which is available in audio format on the Brazilian government's website, cataloged in the library of former presidents' materials. "22-06-2010—Discurso do Presidente da República, Luiz Inácio Lula da Silva, no Ato por Belo Monte e pelo Desenvolvimento da Região do Xingu—Altamira-PA," Biblioteca Presidência da República, 2010, http://www.biblioteca.presidencia.gov.br/presidencia/ex-presidentes /luiz-inacio-lula-da-silva/audios/2010-audios-lula/22-06-2010-discurso-do-presidente -da-republica-luiz-inacio-lula-da-silva-no-ato-por-belo-monte-e-pelo-desenvolvimento-da -regiao-do-xingu-.
2. For more on President Lula's conservative economic policies in the first years of his presidency, including critiques of his embrace of free-market policies, see James Petras and Henry Veltmeyer, "Whither Lula's Brazil? Neoliberalism and 'Third Way' Ideology," *Journal of Peasant Studies* 31, no. 1 (October 2003): 1–44; Alvaro Bianchi and Ruy Braga, "Brazil: The Lula Government and Financial Globalization," *Social Forces* 83, no. 4 (June 2005): 1745–62; Edmund Amann and Werner Baer, "The Macroeconomic Record of the Lula Administration, the Roots of Brazil's Inequality, and Attempts to Overcome Them," in *Brazil under Lula*, ed. Joseph L. Love and Werner Baer (New York: Palgrave Macmillan, 2009), 27–43; Armando Castelar Pinheiro, "Two Decades of Privatization in Brazil," in *The Economies of Argentina and Brazil: A Comparative Perspective*, ed. Werner Baer and David Fleischer (Northampton, MA: Edward Elgar Publishing, 2011).
3. For more on how the Workers' Party approach to development policy adapted past approaches for a more market-oriented and globalized economy, see Kathryn Hochstetler and Alfred P. Montero, "The Renewed Developmental State: The National Development Bank and the Brazil Model," *Journal of Development Studies* 49, no. 11 (July 17, 2013): 1–16.
4. Hochstetler and Montero, "The Renewed Developmental State."
5. Eve Z. Bratman, "Contradictions of Green Development: Human Rights and Environmental Norms in Light of Belo Monte Dam Activism," *Journal of Latin American Studies* 46, no. 2 (May 2014): 261–89.
6. For another description of the tensions between local movements and MAB in the context of Belo Monte, see Eve Bratman, "Passive Revolution in the Green Economy: Activism and the Belo Monte Dam," *International Environmental Agreements: Politics, Law and Economics* 15, no. 1 (2015): 61–77.
7. Aniseh S. Bro, Emilio Moran, and Miquéias Freitas Calvi, "Market Participation in the Age of Big Dams: The Belo Monte Hydroelectric Dam and Its Impact on Rural Agrarian Households," *Sustainability (Switzerland)* 10, no. 5 (2018).
8. The value of compensation payments related to hydroelectric projects is determined by the amount of energy produced and the value of that energy, as determined by the national energy agency. Each impacted area receives a specific percentage, which is determined by Brazilian law. See Agência Nacional de Energia Elétrica, "Compensação Financeira pela Utilização de Recursos Hídricos," n.d., http://www2.aneel.gov.br/aplicacoes/cmpf/gerencial/; Agência Nacional de Energia, "A Compensação Financeira e o Seu Município" (Brasília, 2007).
9. The quote is from Shalini Randeria, "The State of Globalization: Legal Plurality, Overlapping Sovereignties and Ambiguous Alliances between Civil Society and the Cunning State in India," *Theory, Culture and Society* 24, no. 1 (2007): 3. Randeria uses and develops her concept

of the cunning state in other articles. See, for example, Shalini Randeria, "Glocalization of Law: Environmental Justice, World Bank, NGOs and the Cunning State in India," *Current Sociology* 51, no. 3 (2003): 305–28; Shalini Randeria and Ciara Grunder, "The (Un)Making of Policy in the Shadow of the World Bank: Infrastructure Development, Urban Resettlement and the Cunning State in India," in *Policy Worlds: Anthropology and the Analysis of Contemporary Power*, ed. Cris Shore, Susan Wright, and Davide Però (New York: Berghahn Books, 2010), 187–204.

10. For details on how the administrations of President Collor and Cardoso privatized the electricity sector, among others, see Kurt Weyland, "The Brazilian State in the New Democracy," *Journal of Interamerican Studies and World Affairs* 39, no. 4 (1998): 63–94; Peter Kingstone, "Critical Issues in Brazil's Energy Sector: The Long (and Uncertain) March to Energy Privatization in Brazil," James A. Baker III Institute for Public Policy of Rice University, March, 2004; https://www.bakerinstitute.org/research/the-long-and-uncertain-march-to-energy-privatization-in-brazil/ Adilson de Oliveira, "Political Economy of the Brazilian Power Industry Reform," in *The Political Economy of Power Sector Reform: The Experiences of Five Major Developing Countries*, ed. David G. Victor and Thomas C. Heller (New York: Cambridge University Press, 2007), 31–75.

11. "Sobre o PAC," accessed June 27, 2019, http://www.pac.gov.br/sobre-o-pac.

12. Ministério Público Federal, "Fase 49ª da Lava Jato Apura Ilícitos na Construção da Usina Hidrelétrica de Belo Monte," 2018, http://www.mpf.mp.br/pr/sala-de-imprensa/noticias-pr/49a-fase-da-lava-jato-apura-ilicitos-na-construcao-da-usina-hidreletrica-de-belo-monte.

13. Kathryn Hochstetler, "The Politics of Environmental Licensing: Energy Projects of the Past and Future in Brazil," *Studies in Comparative International Development* 46 (October 5, 2011): 353. See also Oliveira, "Political Economy of the Brazilian Power Industry Reform."

14. Evelina Dagnino, "Citizenship: A Perverse Confluence," *Development in Practice* 17, no. 4/5 (2007): 549–56.

15. For a clear and concise definition and explanation of neoliberalism as a government approach and political movement, see Johanna Bockman, "Neoliberalism," *Contexts* 12, no. 3 (2013): 14–15.

16. Scholars have examined how states embed themselves within society in order to generate enough approval for development initiatives to move forward without repressive tactics. For scholarship that discusses state-society coalitions and state embeddedness in Brazil, see Peter B. Evans, *Embedded Autonomy: States and Industrial Transformation* (Princeton, NJ: Princeton University Press, 1995); Thomas K. Rudel, "How Do People Transform Landscapes? A Sociological Perspective on Suburban Sprawl and Tropical Deforestation," *American Journal of Sociology* 115, no. 1 (2009): 129–54; Kathryn Hochstetler and Margaret E. Keck, *Greening Brazil: Environmental Activism in State and Society* (Durham, NC: Duke University Press, 2007); Hochstetler, "The Politics of Environmental Licensing." For another description of how state embeddedness and coalitions between the state and society facilitated the construction of Belo Monte, see Bratman, "Passive Revolution in the Green Economy."

17. For more on the development and evolution of environmental impact assessments in Brazil, see Stephanie N. T. Landim and Luis E. Sánchez, "The Contents and Scope of Environmental Impact Statements: How Do They Evolve over Time?," *Impact Assessment and Project Appraisal* 30, no. 4 (December 2012): 217–28; Luis E. Sánchez, "Development of Environmental Impact Assessment in Brazil," *Schwerpunktthema* 27, no. 4–5 (2013): 193–200.

18. Sabrina McCormick, "The Brazilian Anti-Dam Movement: Knowledge Contestation as Communicative Action," *Organization and Environment* 19, no. 3 (September 1, 2006): 321–46; Hochstetler, "The Politics of Environmental Licensing."

19. For details of the organization and publication timeline of the environmental impact assessment and *RIMA*, see Sônia Barbosa Magalhães and Francisco del Moral Hernández,

eds., *Painel de Especialistas: Análise Crítica do Estudo de Impacto Ambiental do Aproveitamento Hidrelétrico de Belo Monte*, (Belém, Pará, 2009): 12-13, 36-38. For critiques of the public hearings, see Rodolfo Salm, "Belo Monte: A Farsa das Audiências Públicas," EcoDebate, 2009, https://www.ecodebate.com.br/2009/10/08/belo-monte-a-farsa-das-audiencias-publicas-artigo-de-rodolfo-salm/; Philip Martin Fearnside, "Belo Monte: Lições da Luta 11—A Farsa da Audiência Pública," Amazônia Real, 2018, https://amazoniareal.com.br/belo-monte-licoes-da-luta-11-farsa-da-audiencia-publica/; Thiago Almeida Barros and Nírvia Ravena, "Representações Socias nas Audiências Públicas de Belo Monte: Do Palco Ao Recorte Midiático," in *Compolítica—Associação Brasileira de Pesquisadores Em Comunicação e Política* (Rio de Janeiro, 2011).

20. Sônia Barbosa Magalhães and Francisco del Moral Hernández, eds., *Painel de Especialistas*.

21. Norte Energia, *PBA: Projeto Básico Ambiental—Versão Final*, 2011.

22. Lesley K. McAllister, *Making Law Matter: Environmental Protection and Legal Institutions in Brazil* (Stanford, CA: Stanford University Press, 2008). For more on the organization and role of the public prosecutors in Brazil, see Fábio Kerche, "O Ministério Público e a Constituinte de 1987/88," in *O Sistema de Justiça*, ed. Maria Tereza Sadek (Rio de Janeiro: Centro Edelstein de Pesquisas Sociais, 2010), 106–37; Salo Vinocur Coslovsky, "Relational Regulation in the Brazilian Ministério Publico: The Organizational Basis of Regulatory Responsiveness," *Regulation and Governance* 5, no. 1 (2011): 70–89; Hugo Nigro Mazzilli, *Regime Jurídico do Ministério Público*, 7th ed. (São Paulo: Editora Saraiva, 2013).

23. For a list of the MPF's legal processes about Belo Monte, see Ministério Público Federal, "Tabela Belo Monte—Procuradoria Regional da República da 1ª Região," accessed October 27, 2019, http://www.mpf.mp.br/regiao1/sala-de-imprensa/docs/tabela-belo-monte/view.

24. Instituto de Pesquisa Econômica Aplicada, "Mapa da Defensoria: Defensores nos Estados," accessed November 16, 2019, http://www.ipea.gov.br/sites/en-GB/mapadefensoria/defensoresnosestados.

25. Defensoria Pública do Pará, "Defensoria Reinstitui Grupo Especial de Trabalho de Belo Monte," Jusbrasil, 2012, https://dp-pa.jusbrasil.com.br/noticias/3140463/defensoria-reinstitui-grupo-especial-de-trabalho-de-belo-monte.

26. Bibiana Graeff, "Should We Adopt a Specific Regulation to Protect People That Are Displaced by Hydroelectric Projects? Reflections Based on Brazilian Law and the 'Belo Monte' Case," *Florida A&M University Law Review* 7, no. 2 (2012): 261–89; Bratman, "Contradictions of Green Development."

27. Antonio Gramsci, *Selections from the Prison Notebooks of Antonio Gramsci*, ed. Quintin Hoare and Geoffrey Nowell-Smith (New York: International Publishers, 1971).

4. THE LIVING PROCESS

1. Issues of representation are central concerns in the literature on deliberative democracy, given that direct engagement—as opposed to representative democracy—is precisely the argument for instituting participatory spaces. Much of the scholarship lauds the opportunities that new forms of participation offer citizens. See Archon Fung and Erik Olin Wright, *Deepening Democracy: Institutional Innovations in Empowered Participatory Governance* (London: Verso, 2003); Nicole Curato, John S. Dryzek, Selen A. Ercan, Carolyn M. Hendriks, and Simon Niemeyer, "Twelve Key Findings in Deliberative Democracy Research," *Daedalus* 146, no. 3 (2017): 28–38. Other scholars have shown that novel forms of democracy have created challenges around political representation, in that civil organizations—and civil society, more broadly—lack the mechanisms to determine who has the legitimacy and authority to represent whom. See Peter Houtzager and Adrian Gurza Lavalle, "Civil Society's Claims to

Political Representation in Brazil," *Studies in Comparative International Development* 45, no. 1 (2010): 1–29.

2. Presidência da República—Casa Civil, Decreto 7.340, de 21.10.2010.

3. These understandings of participatory and deliberative democracy come from the scholarship on deepening democracy more broadly. For a summary of this literature, see John Gaventa, "Triumph, Deficit or Contestation? Deepening the 'Deepening Democracy' Debate," *IDS Working Paper*, no. 264 (July 2006): 1–34. For a discussion of the debate over whether participation and deliberation work together or in contradiction, see Carole Pateman, "Participatory Democracy Revisited," *Perspectives on Politics* 10, no. 1 (March 2012): 7–19; Curato et al., "Twelve Key Findings in Deliberative Democracy Research." Also see John Parkinson and Jane Mansbridge, eds., *Deliberative Systems: Deliberative Democracy at the Large Scale* (Cambridge: Cambridge University Press, 2012); D. C. Mutz, *Hearing the Other Side: Deliberative versus Participatory Democracy* (Cambridge: Cambridge University Press, 2006).

4. Fung and Wright, *Deepening Democracy*, 5.

5. Scholars and practitioners have debated the qualities that make an institution, a community, or a person empowered. The World Bank has described empowerment as the ability to make choices and turn them into actions and outcomes. Ruth Alsop, Mette Frost Bertelsen, and Jeremy Holland, *Empowerment in Practice: From Analysis to Implementation* (Washington, DC: World Bank, 2006), 10. Scholarship on school choice offers a clear distinction of weak versus strong empowerment. In that literature, decision-making and the ability to make choices are considered weak empowerment. Strong forms of empowerment provide people with access to political processes and responsive political actors and institutions. See Mary Pattillo, "Everyday Politics of School Choice in the Black Community," *Du Bois Review: Social Science Research on Race* 12, no. 1 (2015): 41–71.

6. This analysis builds on the argument made by me and the other authors of *The Civic Imagination*. In that case, Americans used rhetoric of political disavowal that moved them away from terrain they saw as polluted, in order to engage in civic life and to rescue notions of a better democracy. In the case of CGDEX, that argument is turned on its head. Gianpaolo Baiocchi, Elizabeth A. Bennett, Alissa Cordner, Peter Taylor Klein, and Stephanie Savell, *The Civic Imagination: Making a Difference in American Political Life* (Boulder, CO: Paradigm Publishers, 2014); Elizabeth A. Bennett, Alissa Cordner, Peter Taylor Klein, Stephanie Savell, and Gianpaolo Baiocchi, "Disavowing Politics: Civic Engagement in an Era of Political Skepticism," *American Journal of Sociology* 119, no. 2 (September 2013): 518–48.

7. For other cases in which activists use storytelling and theater to make claims, see Julie Sze, *Environmental Justice in a Moment of Danger* (Oakland: University of California Press, 2020).

8. Traditional conceptions of human capital focus on how skills and attributes allow people to perform labor and produce economic value. For this understanding of human capital in economic terms, see Theodore W. Schultz, "Investment in Human Capital," *American Economic Review* 51, no. 1 (1961): 1–17. I extend this idea of human capital, in that the same skills, knowledge, and other characteristics that can be used to perform labor could be used to engage in participatory and deliberative processes.

9. The levels of education are from the Altamira microregion, a government-designated area made up of eight municipalities, all of which are officially recognized as affected by Belo Monte. The statistics come from the 2010 Brazilian Census, "Censo Demográfico 2010—Educação—Resultados da Amostra," Instituto Brasileiro de Geografia e Estatística, accessed August 7, 2019, https://www.ibge.gov.br/estatisticas/sociais/educacao/9662-censo-demografico-2010.html?=&t=downloads.

10. The idea that participants were using small strategic actions as a meaningful and significant form of struggle for power builds on a body of scholarship that examines everyday resis-

tance, which emerged from James C. Scott, *Weapons of the Weak: Everyday Forms of Peasant Resistance* (New Haven, CT: Yale University Press, 1985).

11. For other critical scholarship on the unintended consequences of participatory institutions, including both the entrenchment of inequalities and the creation of new opportunities for participation that can emerge out of such institutions, see Caroline W. Lee, Michael McQuarrie, and Edward T. Walker, eds., *Democratizing Inequalities: Dilemmas of the New Public Participation* (New York: New York University Press, 2015).

5. THE FIGHT FOR RECOGNITION

1. The language of "mobilizing" and "activating" the state draws on the scholarship of Rebecca Abers and Margaret Keck, who show how activists both within and outside of the state work together to marshal the capacity of the state to act on behalf of the public interest. Rebecca Neaera Abers and Margaret E. Keck, "Mobilizing the State: The Erratic Partner in Brazil's Participatory Water Policy," *Politics and Society* 37, no. 2 (April 3, 2009): 289–314.

2. As outlined in previous chapters, the *Summary of the Environmental Impact and the Basic Environmental Project* outlined Belo Monte's impacts and the plans to address those impacts, but scholars and activists contested these assessments. For the official documents, see Eletrobrás, *RIMA—Relatório de Impacto Ambiental—Aproveitamento Hidrelétrico Belo Monte*, 2009; Norte Energia, *PBA: Projeto Básico Ambiental—Versão Final, Versão Final*, 2011. For examples of scholarship and documents that challenge the official assessments, see Oswaldo Sevá Filho, ed., *Tenotã- Mõ: Alertas sobre as Conseqüências dos Projetos Hidrelétricos no Rio Xingu* (International Rivers Network, 2005); Sônia Barbosa Magalhães and Francisco del Moral Hernández, eds., *Painel de Especialistas: Análise Crítica do Estudo de Impacto Ambiental do Aproveitamento Hidrelétrico de Belo Monte*, (Belém, 2009).

3. Studies have documented diverse and endemic plant and animal species in the Volta Grande. Prior to construction, scientists suggested Belo Monte could alter the Volta Grande's ecosystem, and studies carried out during construction showed ongoing concerns. For examples, see Magalhães and Hernández, *Painel de Especialistas*; A. O. Sawakuchi, G. A. Hartmann, H. O. Sawakuchi, F. N. Pupim, D. J. Bertassoli, M. Parra, J. L. Antinao, L. M. Sousa, M. H. Sabaj Pérez, P. E. Oliveira, R. A. Santos, J. F. Savian, C. H. Grohmann, V. B. Medeiros, Michael M. McGlue, D. C. Bicudo, and S. B. Faustino, "The Volta Grande Do Xingu: Reconstruction of Past Environments and Forecasting of Future Scenarios of a Unique Amazonian Fluvial Landscape," *Scientific Drilling* 20 (2015): 21–32; Daniel B. Fitzgerald, Mark H. Sabaj Pérez, Leandro M. Sousa, Alany P. Gonçalves, Lucia Rapp Py-Daniel, Nathan K. Lujan, Jansen Zuanon, Kirk O. Winemiller, and John G. Lundberg, "Diversity and Community Structure of Rapids-Dwelling Fishes of the Xingu River: Implications for Conservation amid Large-Scale Hydroelectric Development," *Biological Conservation* 222 (June 2018): 104–12.

4. In addition to the concerns expressed by Paulo and his colleagues, scientists have also documented numerous other ways the highly diverse fish species have adapted to the unique habitats in the Volta Grande of the Xingu River. The series of rapids, braided channels, and small islands changed with the construction of Belo Monte, threatening these fish species. Fitzgerald et al., "Diversity and Community Structure of Rapids-Dwelling Fishes of the Xingu River."

5. For reporting of the first protest carried out by fishers, see Fátima Lessa, "Pescadores Realizam Manifestação Contra Belo Monte," *EXAME*, March 14, 2011.

6. Social movement scholarship shows how collective action frames are developed through shared and evolving processes, help shape group meaning, and challenge existing understandings. The literature on framing originates from Erving Goffman's work and has been used

extensively to interpret and assess collective action. Erving Goffman, *Frame Analysis: An Essay on the Organization of Experience* (Cambridge, MA: Harvard University Press, 1974); Robert D. Benford and David A. Snow, "Framing Processes and Social Movements: An Overview and Assessment," *Annual Reveiw of Sociology* 26 (2000): 611–39.

7. "Inaugurada Casa de Governo Em Altamira (PA)," Ministério do Planejamento, 2013.

8. Randeria uses the term "cunning state" to highlight the strategies that the state employs, rather than capacities that the state is presumed to have or lack. See Shalini Randeria, "Glocalization of Law: Environmental Justice, World Bank, NGOs and the Cunning State in India," *Current Sociology* 51, no. 3 (2003): 305–28.

9. For more on the role of mediation and translation in civil society and the public sphere, see Gianpaolo Baiocchi, Patrick Heller, and Marcelo Silva, *Bootstrapping Democracy: Transforming Local Governance and Civil Society in Brazil* (Stanford, CA: Stanford University Press, 2011), 31–33.

10. Wendy Wolford insightfully describes how, in the absence of a strong state, social movements played a mediating role in Brazilian land reform. She argues that participatory democratic processes that were not planned—and existed neither in policy nor in theory—nevertheless developed "by default rather than by design" as a result of social movements transgressing acceptable and legal boundaries of action. Wendy Wolford, "Participatory Democracy by Default: Land Reform, Social Movements and the State in Brazil," *Journal of Peasant Studies* 37, no. 1 (2010): 91–109. Rebecca Abers and Margaret Keck similarly argue that a weak state can be "activated" by civic society through participatory spaces. Abers and Keck, "Mobilizing the State."

11. While the concept of collective identities is used in a broad range of scholarship, the definition here draws on the ways it has been used in the literature on social movements and also borrows from José Itzigsohn's work on Dominican immigrants, specifically his description of categorical collective identities. Francesca Polletta and James M. Jasper, "Collective Identity and Social Movements," *Annual Reveiw of Sociology* 27 (2001): 283–305; José Itzigsohn, *Encountering American Faultlines: Race, Class, and the Dominican Experience in Providence* (New York: Russell Sage Foundation, 2009), 119–20.

12. The attention to understanding aligns with the field of political epistemology, which focuses on "politics-oriented knowledge-making practices of people and their consequences." Andreas Glaeser, *Political Epistemics: The Secret Police, the Opposition, and the End of East German Socialism* (Chicago: University of Chicago Press, 2011), xxvi. My analysis also builds on the work of scholars who examine how expertise and knowledge are created, used, and contested, particularly in environmental conflicts. See Sabrina McCormick, *Mobilizing Science: Movements, Participation, and the Remaking of Knowledge* (Philadelphia: Temple University Press, 2009); Fabiana Li, *Unearthing Conflict: Corporate Mining, Activism, and Expertise in Peru* (Durham, NC: Duke University Press, 2015); Javiera Barandiarán, *Science and Environment in Chile: The Politics of Expert Advice in a Neoliberal Democracy* (Cambridge, MA: MIT Press, 2018).

13. Ana de Francesco and Cristiane Carneiro, eds., *Atlas dos Impactos da UHE Belo Monte sobre a Pesca* (São Paulo: Instituto Socioambiental, 2015), 5.

14. Leinad Ayer O. Santos and Lúcia M. M. de Andrade, eds., *As Hidrelétricas do Xingu e os Povos Indígenas* (São Paulo: Comissão Pró-Indio de São Paulo, 1988).

15. Filho, *Tenotã-Mõ: Alertas sobre as Conseqüências dos Projetos Hidrelétricos No Rio Xingu*, 7.

16. Magalhães and Hernández, *Painel de Especialistas*.

17. For more details on public prosecutors in Brazil that tend to partner with other organizations and actors outside of the office, see Salo Coslovsky's discussion of what he calls "reformist" prosecutors that act in slightly deviant ways in order to solve problems. Salo Vinocur Coslovsky, "Relational Regulation in the Brazilian Ministério Publico: The Organizational

Basis of Regulatory Responsiveness," *Regulation and Governance* 5, no. 1 (2011): 70–89. Also see Cátia Silva's description of what she calls "fact-oriented" prosecutors. Cátia Aida Silva, "Promotores de Justiça e Novas Formas de Atuação Em Defesa de Interesses Sociais e Coletivos," *Revista Brasileira de Ciências Sociais* 16, no. 45 (2001): 127–44.

18. Brown details how communities have effectively used citizen-science alliances—defined as "lay-professional collaborations in which citizens and scientists work together on issues identified by laypeople"—to seek compensation for health impacts due to corporate pollution. Phil Brown, *Toxic Exposures: Contested Illnesses and the Environmental Health Movement* (New York: Columbia University Press, 2007), 33.

19. The timeline of events depicted in this section come from discussions with people involved, as well as from the book referenced in the text. Sônia Barbosa Magalhães and Manuea Carneiro da Cunha, eds., *A Expulsão de Ribeirinhos em Belo Monte: Relatório da SBPC* (São Paulo: Sociedade Brasileira para o Progresso da Ciência—SBPC, 2017).

6. THE LAW, ACTIVISM, AND LEGITIMACY

1. For foundational texts on legal mobilization, see Stuart A. Scheingold, *The Politics of Rights: Lawyers, Public Policy, and Political Change, The Politics of Rights*, 2nd ed. (Ann Arbor: University of Michigan Press, 2004); Boaventura de Sousa Santos Santos and César A. Rodríguez-Garavito, eds., *Law and Globalization from Below: Towards a Cosmopolitan Legality* (Cambridge: Cambridge University Press, 2005); Michael W. McCann, ed., *Law and Social Movements*, 2nd ed. (New York: Routledge, 2017).

2. Marc Galanter, "Why the 'Haves' Come Out Ahead: Speculations on the Limits of Legal Change," *Law and Society Review* 9, no. 1 (1974): 95; David Kairys, *The Politics of Law: A Progressive Critique*, 3rd ed. (Philadelphia: Basic Books, 1998); Gerald N. Rosenberg, *The Hollow Hope: Can Courts Bring About Social Change?*, 2nd ed. (Chicago: University of Chicago Press, 2008).

3. For examples of edited volumes and reviews that detail debates on whether the law and the use of lawyers support or degrade social movements, how litigation impacts social movements beyond the outcomes of particular cases, and the ways the use of the courts influences broader political contexts, see Holly J. McCammon and Allison R. McGrath, "Litigating Change? Social Movements and the Court System," *Sociology Compass* 9, no. 2 (February 1, 2015): 128–39; Austin Sarat and Stuart A. Scheingold, eds., *Cause Lawyers and Social Movements* (Stanford, CA: Stanford University Press, 2006); Anna-Maria Marshall and Daniel Crocker Hale, "Cause Lawyering," *Annual Review of Law and Social Science* 10, no. 1 (November 3, 2014): 301–20; Sandra R. Levitsky, "Law and Social Movements: Old Debates and New Directions," in *The Handbook of Law and Society*, ed. Austin Sarat and Patricia Ewick (Malden, MA: John Wiley and Sons, 2015), 382–98; Steven A. Boutcher, "Law and Social Movements: It's More than Just Litigation and Courts," *Mobilizing Ideas*, February 2018, https://mobilizingideas.wordpress.com/2013/02/18/law-and-social-movements-its-more-than-just-litigation-and-courts/.

4. Kathryn Hochstetler and Margaret E. Keck, *Greening Brazil: Environmental Activism in State and Society* (Durham, NC: Duke University Press, 2007); Lesley K. McAllister, *Making Law Matter: Environmental Protection and Legal Institutions in Brazil* (Stanford, CA: Stanford University Press, 2008).

5. For an account of why and how more Indigenous people in Brazil began claiming their "Indianness" at the end of the twentieth century, see Jonathan W. Warren, *Racial Revolutions: Antiracism and Indian Resurgence in Brazil* (Durham, NC: Duke University Press, 2001).

6. For examples of scholarship that detail the impacts of supportive relations between legal institutions, lawyers, and activists, see Michael W. McCann, *Rights at Work: Pay Equity Reform*

and the Politics of Legal Mobilization (Chicago: University of Chicago Press, 1994); Austin Sarat and Stuart A. Scheingold, eds., *The Worlds Cause Lawyers Make: Structure and Agency in Legal Practice* (Stanford, CA: Stanford University Press, 2005). For a discussion of how public defenders and public prosecutors in Brazil are entangled with a wide assortment of actors in society, see Salo Vinocur Coslovsky, "Relational Regulation in the Brazilian Ministério Publico: The Organizational Basis of Regulatory Responsiveness," *Regulation and Governance* 5, no. 1 (2011): 70–89; Salo Vinocur Coslovsky, "Beyond Bureaucracy: How Prosecutors and Public Defenders Enforce Urban Planning Laws in São Paulo, Brazil," *International Journal of Urban and Regional Research* 39, no. 6 (2015): 1103–19.

7. Lucie E. White, "Mobilization on the Margins of the Lawsuit: Making Space for Clients to Speak," *New York University Review of Law and Social Change* 16, no. 4 (1987): 535–64; Austin Sarat and Stuart Scheingold, "Cause Lawyering and the Reproduction of Professional Authority: An Introduction," in *Cause Lawyering: Political Commitments and Professional Responsibilities*, ed. Austin Sarat and Stuart A. Scheingold (Oxford: Oxford University Press, 1998), 3–28; Scheingold, *The Politics of Rights*.

8. Some cause lawyers have responded to critiques by blending their professional identity with their activist identity or by fabricating identities as they adapt to situational and institutional limitations. For an example of the ways social movement literature on identity work can be used to understand how lawyers have navigated these boundaries, see Lynn C. Jones, "Exploring the Sources of Cause and Career Correspondence among Cause Lawyers," in Sarat and Scheingold, *The Worlds Cause Lawyers Make*, 203–38. Other lawyers have built creative legal challenges in ways that make marginalized communities more visible. See Scott Barclay and Anna-Maria Marshall, "Supporting a Cause, Developing a Movement, and Consolidating a Practice: Cause Lawyers and Sexual Orientation Litigation in Vermont," in Sarat and Scheingold, *The Worlds Cause Lawyers Make*, 171–202.

9. For other examples in which public prosecutors—and public defenders, to some extent—in Brazil have used creative and diverse strategies to support particular causes, see McAllister, *Making Law Matter*; Lesley K. McAllister, "Environmental Advocacy Litigation in Brazil and the United States," *Journal of Comparative Law* 6 (2011): 2; Coslovsky, "Relational Regulation in the Brazilian Ministério Publico"; Coslovsky, "Beyond Bureaucracy."

10. Political theorists have long debated what is meant by "the political." Some, like Hannah Arendt and Jürgen Habermas, view it as that which is debated and deliberated, while others, like Chantal Mouffe, argue it is that which is based on power, conflict, and antagonism, which is part and parcel of human society. Chantal Mouffe, *On the Political: Thinking in Action* (New York: Routledge, 2005).

11. David Mosse, *Cultivating Development: An Ethnography of Aid Policy and Practice* (London: Pluto Press, 2005), 107.

12. McCann, *Rights at Work*.

CONCLUSION

1. In pointing out the underlying meanings of "democracy" and "development," Radha D'Souza highlights the conceptual contradictions of democratic development, arguing for a reconceptualizing of the development paradigm. Radha D'Souza, "The Democracy-Development Tension in Dam Projects: The Long Hand of the Law," *Political Geography* 23, no. 6 (August 2004): 701–30. For more on the concept of democratic developmentalism, see Mark Robinson and Gordon White, eds., *The Democratic Developmental State: Politics and Institutional Design* (New York: Oxford University Press, 1998).

2. For some examples of Brazil's participatory processes, see Rebecca Abers, "From Ideas to Practice: The Partido dos Trabalhadores and Participatory Governance in Brazil," *Latin American Perspectives* 23, no. 4 (1996): 35–53; Gianpaolo Baiocchi, *Militants and Citizens: The Politics of Participatory Democracy in Porto Alegre* (Stanford, CA: Stanford University Press, 2005); Brian Wampler, *Participatory Budgeting in Brazil: Contestation, Cooperation, and Accountability* (University Park: Pennsylvania University Press, 2007); Nathan Engle, Maria Carmen Lemos, Lori Kumler, and Rebecca Abers, "The Watermark Project: Analyzing Water Management Reform in Brazil," *Journal of the International Institute* 13, no. 2 (2006); Leonardo Avritzer, *Participatory Institutions in Democratic Brazil* (Baltimore, MD: Johns Hopkins University Press, 2009); Gianpaolo Baiocchi, Patrick Heller, and Marcelo Silva, *Bootstrapping Democracy: Transforming Local Governance and Civil Society in Brazil* (Stanford, CA: Stanford University Press, 2011); Christopher L. Gibson, *Movement-Driven Development: The Politics of Health and Democracy in Brazil* (Stanford, CA: Stanford University Press, 2019).

3. The Intergovernmental Panel on Climate Change released its fifth comprehensive assessment report in 2014 and additional reports from working groups since that time. All of these documents suggest the planet will continue to warm and the impacts will increase. IPCC, "Climate Change 2014: Synthesis Report. Contributions of Working Groups I, II, and III to the Fifth Assessment Report of the Intergovernmental Panel on Climate Change" (Geneva, Switzerland, 2014). See the IPCC website for reports: https://www.ipcc.ch/.

4. A great deal of scholarship has been published on the differing impacts of climate change and disasters along lines of race, class, gender, and other social markers. For examples, see Geraldine Terry, "No Climate Justice without Gender Justice: An Overview of the Issues," *Gender and Development* 17, no. 1 (2009): 5–18; Sharon L. Harlan, David N. Pellow, J. Timmons Roberts, Shannon Elizabeth Bell, William G. Holt, and Joane Nagel, "Climate Justice and Inequality," in *Climate Change and Society: Sociological Perspectives*, ed. Riley E. Dunlap and Robert J. Brulle (New York: Oxford University Press, 2015), 127-163. Bob Bolin and Liza C. Kurtz, "Race, Class, Ethnicity, and Disaster Vulnerability," in *Handbook of Disaster Research*, ed. Havidán Rodríguez, William Donner, and Joseph E. Trainor, 2nd ed. (Cham: Springer, 2018), 181–203. For a moral and religious argument for addressing climate change, see Pope Francis's encyclical letter. Pope Francis, *Encyclical on Climate Change and Inequality: On Care for Our Common Home* (New York: Melville House, 2015).

5. For details on the environmental justice framework as it developed in the 1980s and early 1990s, see Robert D. Bullard, "Environmental Justice for All," in *Unequal Protection: Environmental Justice and Communities of Color*, ed. Robert D. Bullard (San Fransisco: Sierra Club Books, 1994), 3–22. The definition of environmental justice provided here also closely resembles that of the U.S. EPA. United States Environmental Protection Agency, "Environmental Justice," https://www.epa.gov/environmentaljustice. For early examples of noteworthy cases, see Lois Marie Gibbs and Murray Levine, *Love Canal: My Story* (Albany: State University of New York Press, 1982); Robert D. Bullard, *Dumping in Dixie: Race, Class, and Environmental Quality*, 3rd ed. (Boulder, CO: Westview Press, 2000). For more recent trends in scholarship on climate justice, environmental justice, and a just transition, see David Ciplet, J. Timmons Roberts, and Mizan R. Khan, *Power in a Warming World: The New Global Politics of Climate Change and the Remaking of Environmental Inequality* (Cambridge, MA: MIT Press, 2015); Julian Agyeman, David Schlosberg, Luke Craven, and Caitlin Matthews, "Trends and Directions in Environmental Justice: From Inequity to Everyday Life, Community, and Just Sustainabilities," *Annual Review of Environment and Resources* 41, no. 1 (2016): 321–40; Julie Sze, *Environmental Justice in a Moment of Danger* (Oakland: University of California Press, 2020). For a theoretical approach using a "just adaptation" framework, see David Schlosberg,

"Climate Justice and Capabilities: A Framework for Adaptation Policy," *Ethics and International Affairs* 26, no. 4 (2012): 445–61.

6. Ed Atkins, *Contesting Hydropower in the Brazilian Amazon* (New York: Routledge, 2021).

7. The *Guardian* and *El País* first reported on Norte Energia's letter to the federal government. Their articles also explain how climate change affects the dam's structural integrity. Eliane Brum, "Erro de Projeto Coloca Estrutura de Belo Monte em Risco," *El País Brasil*, November 8, 2019; Jonathan Watts, "Poorly Planned Amazon Dam Project 'Poses Serious Threat to Life,'" *Guardian*, November 8, 2019. For Norte Energia's response to these reports, see "Estrutura de Belo Monte Não Corre Risco, Diz Norte Energia," *El País Brasil*, November 21, 2019.

8. For other examples in which citizens mediated state functions or mobilized parts of the state apparatus, see Rebecca Neaera Abers and Margaret E. Keck, "Mobilizing the State: The Erratic Partner in Brazil's Participatory Water Policy," *Politics and Society* 37, no. 2 (April 3, 2009): 289–314; Wendy Wolford, "Participatory Democracy by Default: Land Reform, Social Movements and the State in Brazil," *Journal of Peasant Studies* 37, no. 1 (2010): 91–109.

9. Sabrina McCormick, "Democratizing Science Movements," *Social Studies of Science* 37, no. 4 (August 29, 2007): 609–23; Sabrina McCormick, *Mobilizing Science: Movements, Participation, and the Remaking of Knowledge* (Philadelphia: Temple University Press, 2009).

10. Phil Brown, *Toxic Exposures: Contested Illnesses and the Environmental Health Movement* (New York: Columbia University Press, 2007).

11. Fabiana Li, *Unearthing Conflict: Corporate Mining, Activism, and Expertise in Peru* (Durham, NC: Duke University Press, 2015); Alissa Cordner, *Toxic Safety: Flame Retardants, Chemical Controversies, and Environmental Health* (New York: Columbia University Press, 2019).

12. Lesley K. McAllister, *Making Law Matter: Environmental Protection and Legal Institutions in Brazil* (Stanford, CA: Stanford University Press, 2008).

13. Dorceta E. Taylor, *The Rise of the American Conservation Movement: Power, Privilege, and Environmental Protection* (Durham, NC: Duke University Press, 2016).

14. Celene Krauss, "Women and Toxic Waste Protests: Race, Class and Gender as Resources of Resistance," *Qualitative Sociology* 16, no. 3 (September 1993): 247–62.

15. For an argument about how hegemonic masculinity hinders men from being involved in environmental justice activism, as well as more details about the ways women's identities shape activism, see Shannon Elizabeth Bell and Yvonne A. Braun, "Coal, Identity, and the Gendering of Environmental Justice Activism in Central Appalachia," *Gender and Society* 24, no. 6 (2010): 794–813. See also Phil Brown and Faith I. T. Ferguson, "'Making a Big Stink': Women's Work, Women's Relationships, and Toxic Waste Activism," *Gender and Society* 9, no. 2 (1995): 145–72.

INDEX

Abers, Rebecca, 209n1, 210n10
abundance: collective action and, 190n6
Acselrad, Henri, 34
activism: *vs.* legal action, 151
agnosticism, 190n8
Alberto (fisher), 125, 126, 129, 135
Altamira: agronomy studies, 94; air pollution, 45; anti-dam movement, 69–70, 72; Catholic Church, 53; CGDEX meetings, 97; civil society consolidation, 58–63; colonization of Amazonia and, 50–51; crime rate, 44, 45, 199n43; dam construction's effect on, 71–72, 76; displacement of residents, 44, 103; economic development, 44–45, 48–49, 51; education, 55, 108, 208n9; electricity in, 61–62; fishers' protest in, 127–28; food prices, 72; housing, 104–5, 158; infrastructure development, 49, 51, 55, 94; lack of human capital, 108; living conditions, 43; political struggles, 14; population, 44, 45, 48–49, 51, 53, 71–72, 199n44, 201n11; public defenders' office, 84, 158, 159; public prosecutors, 83; rent prices, 45; schools, 49, 54–55; sewer system, 45; social programs, 49; unemployment, 158; women's movement in, 177–78; Workers' Party in, 63
aluminum industry, 33, 195n19
Amazonia: colonization of, 47, 49–51, 201n11; dam construction in, 33, 71; natural resources, 33, 48–49, 50, 200n1; participatory initiatives, 41; settlements, 49, 50, 51–52; sustainable development, 203n30
Amazon Watch, 31
American pragmatist school, 17, 191n14
anti-dam activists: meetings, 90; participatory opportunity, 91; slogans, 27; split within, 67–68, 70, 85–86; support network, 58
Appalachia: environmental justice movement, 177
Argentina: hydropower in, 30
Artur (public defender), 6, 140, 141, 142, 149
Atkins, Ed, 170

Babaquara Dam, 37
Baines, Stephen, 196n24
Baiocchi, Gianpaolo, 23
Balbina Dam: energy capacity, 35, 196n23; environmental impact, 34–35, 196n24; planning, 35
barrageiros, 45
Barreto, Andreia Macedo: building of institutional legitimacy, 153, 154; disaffiliation with CGDEX, 76; leadership, 164; legal representation of riverine communities, 1–3, 4, 5, 84, 114, 129–30, 132, 152, 153, 155, 159; mining project protests and, 161, 162–63, 164; mobilization of women, 179; outreach efforts, 149; personality, 140; public speaking skills, 163; relationships with government agencies, 158; reputation of, 176; on role of public defenders, 144, 151–52, 157
Belém, 162
Belo Monte Dam: assessment of impact, 62, 66, 81, 82, 134–35; benefits of, 82, 112, 170; compensation program, 8–9, 44; construction of, 6, 11, 12, 16, 42, 44, 71, 86; cost of, 42; design of, 38, 40; displacement of population, 4, 16, 43, 44; energy capacity of, 38; environmental impact, 1–2, 4, 8–9, 13–14, 28, 44, 45, 114–15, 118–19, 120, 159–60, 171; federal government investments, 19, 42, 64, 67, 78; flood zone, 38; funding, 22; legal challenges, 75–76, 82–85, 159; licensing procedures, 42, 82, 85; main powerhouse dam of, 39; on map, 39; mitigation processes, 42–44, 66, 76–77, 78–79; opposition to, 1–2, 3, 9, 12, 14, 15–16, 71, 73, 85, 151, 162; participatory processes, 22, 23, 42, 43, 44, 75; planning, 11, 37–38, 62; political aspects, 13, 14; press conference at, 27–28; promotion of, 27–28; public and private investments, 66, 78, 79; public debates over, 2, 28, 166–67, 180; public hearings, 81, 82; security concerns, 4; social impact of, 11–12, 13–14, 18, 19, 28, 44–45, 71–73, 115; supporters, 43, 67, 68, 82, 85, 110, 170; transportation challenges, 4; viability studies, 81

Belo Monte Dam negotiations: agreements, 9; ethnographic research, 16–17; FUNAI representative, 6; Indigenous protestors, 4–5, 6; Norte Energia's officials, 4–5, 9; participatory opportunities, 85–86; police delegates, 5; public defenders, 75–76; public prosecutor's office, 82–83; recording of, 4–5; security issues, 5, 6; state officials at, 7–8; translators and mediators, 23–24, 132

Belo Sun Mining: employment opportunities, 161; gold reserves exploration, 150, 159, 163; legal institutions and, 164; licensing, 163; opposition to, 151, 161, 162; supporters of, 162, 163, 164

Big Bend. *See* Volta Grande (Big Bend) region

Black Movement, 60, 178

blocking coalitions, 16

BNDES (National Bank for Economic and Social Development), 58, 67, 83

Boff, Clodovis, 52, 201n13

Boff, Leonardo, 52, 201n13

Bolsa Familia (social program), 41, 198n35

Bolsonaro, Jair, 42, 43, 180

Brazil: abolition of slavery, 48; capital of, 49; colonial history, 47, 48; communication, 51; constitution of, 79; cotton boom, 48; democratic governance, 37; developmentalist paradigm, 32–33, 34, 36, 41, 65–66; energy demands, 170; environmental policies, 58; foreign capital, 36–37, 51; government policies, 35; hydropower, 31–32, 195n14; industrial development, 48, 49; infrastructure projects, 11; military dictatorship, 36–37, 49–51, 180; national debt crisis, 36; neoliberal policies, 79; participatory initiatives, 22, 23, 41–42, 167–68, 198n38; political regime, 32, 34; political scandals, 12–13, 78, 180, 189n3; presidential elections, 59–60; privatization of electricity industry, 77–78; progressive politics, 167; social movements, 12, 35–36; transition to democracy, 56, 59, 179–80; transportation, 51; Workers' Party government, 11, 12, 58, 60, 65

Brown, Phil, 211n18

Bruno (protester), 8

Cardoso, Fernando Henrique, 62, 77–78, 79

Carlindo (activist), 94–95

Carson, Rachel, 177

Carter, Majora, 177

Carvalho, Georgia, 203n29

Catholic Church: opposition to dam projects, 35, 56–57, 58; organizational work of, 54; social mobilization efforts, 46, 49, 52–54, 55, 61, 109; territorial organization, 201n16; women activists in, 178

CCBM (Belo Monte Construction Consortium), 156

CEBs (Christian Base Communities), 52–53, 59

CGDEX (Steering Committee of the Regional Sustainable Development Plan of the Xingu): benefits, 93, 109, 112; bureaucracy, 111; civil society representatives, 110; criticism of, 93–94, 103, 105–6, 111; dam-affected communities and, 86; decision-making power, 42, 43; democratic practices, 79–80, 90, 101, 102, 103, 106, 108–9; fishers representation, 123; formation of, 42, 73, 78; funding, 76, 79, 86, 91–92, 97–98; government use of, 92, 100, 103, 112; housing project and, 104–5; legal services, 75–76; limitations and challenges, 90, 95, 106–7; local groups' engagement with, 73–75, 91, 108; meetings, 89–90, 90, 93, 96–101; members, 92–93, 94, 95; mission, 96, 102; mitigation efforts, 78–80; non-government organizations and, 75–76, 91, 103–4; as participatory space, 73, 76, 85–86, 90–92, 95–96, 100–102, 155, 180–81; positives aspects of, 107–8; problem of voting, 94–95; proponents of, 106, 107, 108; resettlement process and, 103–4; Rural Workers' Union's participation in, 107–10; technical groups and committees, 79, 98–100, **99**, 101; termination of, 43, 180; Xingu Vivo and, 86, 111–12

China: hydropower in, 29, 30

CIMI (Indigenous Missionary Council), 53, 56

Ciplet, David, 19

citizen-science alliances, 173, 211n18

civil society: divisions within, 66, 71, 73

claims-making processes, 79–80, 167, 168, 180

Claudio (leader of FVPP): background, 68; CGDEX and, 74, 79, 97; social activism, 68, 85–86, 111

climate change: adaptation approach to, 17, 18; consequences of, 18, 191n16; hydroelec-

tricity and, 13, 17–18, 171; impact of, 169–70; mitigation approach to, 17; state-led adaptation programs, 13, 18–19; studies of, 213n3
climate injustice. *See* environmental justice
CNBB (National Conference of Brazilian Bishops), 53
CNEC (National Consortium of Engineering Consultants), 37
cocoa industry, 95
collective action, 190n6
collective governance, 76
collective identity, 134, 210n11
Collor de Mello, Fernando Affonso, 77, 79
Columbia River, 30
Commercial, Industrial, and Agropastoral Association (ACIAPA), 96
Commission of Fisheries and Aquaculture, 131
Commission of Fishing, 131
compensation payments: value of, 205n8
CONAMA (National Council on the Environment), 80
Corrêa, Camargo, 34
Coslovsky, Salo Vinocur, 210n17
CRAB (Regional Commission of People Affected by Dams), 35–36
Critical Analysis of the Environmental Impact Assessment of the Belo Monte Hydroelectric Project, 136
Cuiabá-Santarém Highway, 41
Cummings, Barbara, 196n24
"cunning state," 77, 129, 210n8

Dagnino, Evelina, 79
dam-affected communities: demands of, 3–4, 8, 174; government approach to, 34; legal support of, 1, 2, 20; public meetings with, 141–42; resettlement processes, 103; solidarity networks, 12; struggle for rights, 86
dams: assessment of effectiveness of, 193n1; economic benefits, 11, 29, 30; environmental impact, 13, 18, 32, 33–34, 80–81, 171; future of, 166; legal challenges, 65–66; licensing processes, 12, 65, 80–81; local development and, 166; movement against, 35–36, 58; planning stage, 34; political impact, 13; promotion of, 29; purpose of, 28–29; social impact, 13–14, 18, 30–31, 35, 57; as symbol of progress, 29, 30; viability studies for, 62
Dams of the Xingu and the Indigenous People, The, 136
deep democracy, 22, 189n2
deliberation, 167, 168, 169, 181
democracy: effectiveness of, 169; forms of, 207n1; legal agencies and, 181–82; at the local level, 169, 180–81; participation in politics and, 21–22; power and, 20; representation and, 207n1; rule of law and, 182; threat to, 179; *See also* participatory democracy; representative democracy
democratic developmentalism, 11, 13, 38, 80, 166–67, 189n1, 197n34
developmental state, 32, 167, 202n24
Dilermando (fisher), 119, 120
displacement of population, 30–31, 33, 43, 44; *See also* resettlement program
distributive justice, 19
DNAEE (National Department of Water and Electrical Energy), 37
domestic violence, 59
Douglas, Marjory Stoneman, 177
DPE (Public Defenders' Office of the State of Pará), 84, 129–30, 139, 143; *See also* public defenders
DPU (Public Defenders' Office of the Union), 83–84, 85, 153

ecologically unequal exchange, 200n1
Edilson (shopkeeper), 119–20
Eduardo (coordinator for CCBM), 156–57
Egypt: hydropower in, 30
Elena (employee of independent company), 156–57
Eletrobrás company, 37, 56, 58, 62, 66, 81
Eletronorte company, 33–34, 37, 62, 81
Eliana (activist), 98, 101, 179
empowerment: forms of, 208n5
enabling coalitions, 16, 78
environmental impact assessment (EIA), 80–81, 82, 85
environmental inequality, 172
environmental justice, 19, 170, 173, 177, 213n5, 214n15
environmental racism, 177
Expulsion of Riverine People in Belo Monte, The, 137

farmers: blockade of Transamazon highway, 54; displacement of, 35; living conditions, 52, 55; settlement in Amazonia, 49, 51–52, 55; transportation system and, 55
Fearnside, Philip, 33, 196n24, 197n31
Felipe (member of CGDEX), 107
Fernando (MPA representative), 128
fishers: access to river, 122; Belo Monte dam's impact on, 4, 117, 118–21; compensation demands, 8, 116, 138, 180; conciliatory hearing, 129–30; direct actions, 131; displacement of, 122, 133, 134, 153; education of, 121; environmental concerns, 120–21, 124; fishing in prohibited zone, 124, 125, 126, 127–28; government officials and, 128–29, 133; homes of, 121–22, 122; living conditions, 119, 121, 126; marginalized position of, 117; Norte Energia's dialogue with, 117, 123, 124–25, 128–29, 130–33; participatory opportunities, 116–17, 155, 168, 180; protests of, 1–2, 9, 124, 125–26, 127, 127–28; public defenders of, 152–53; social mobilization of, 116, 123–24, 130–31; struggle for justice, 125, 138; support networks, 21, 93, 129, 132, 154, 172–73; union, 118, 123, 124, 134; *See also* MPA (Ministry of Fisheries and Aquaculture); riverine people
FORT Xingu group, 74, 75, 92, 96, 110
Francesco, Ana de, 134–35, 162
FUNAI (National Indian Foundation), 6, 83
Furnas Dam, 32
FVPP (Living, Producing, and Preserving Foundation), 61, 68, 71, 111

Gabi (activist), 95, 108, 112
Gaventa, John, 22
Gayoso, Raymundo José de Souza, 48
Geraldo, Zé, 27, 28, 40
Gibbs, Lois, 177
Gloria (activist), 113–14, 115, 116, 135, 137, 154
Goodall, Jane, 177
Gramsci, Antonio, 20, 21, 86
greenhouse gas emissions: hydroelectric dams and, 18, 34
Greenpeace, 31

hegemony, theory of, 21, 86
Heller, Patrick, 23
Hill, Julia Butterfly, 177

Hochstetler, Kathryn, 16, 78, 202n24, 203n29
Hoover, Herbert, 29
Hoover Dam, 29
human capital, 108, 208n8
hydroelectric dams in Brazil: electricity generated by, 195n14; environmental impact, 18, 34, 35; financing of, 36, 37, 40; foreign interest in, 33; history of, 36; political agenda and, 33, 36, 37, 40; renewable energy and, 17–18, 30; social impact, 13
hyperdeliberation, 23, 167, 169

Ibama (Brazilian Institute of the Environment and Renewable Natural Resources), 81, 82, 83, 120, 132, 136, 137
INCRA (federal land colonization and titling agency), 61
India: hydropower in, 29, 30, 31
Indigenous communities: compensation to, 8, 94, 146, 151; demarcated territories, 53, 146; displacement of, 35, 196n24; legal counselling, 4; political organization of, 56; protests against Belo Monte Dam, 1–2, 15, 56–57, 57, 104; publications of, 136; struggle for environmental justice, 172; support groups, 53
infrastructure projects: as developmental schemes, 166–67; environmental impact, 80; financing of, 202n25; global actors, 21; licensing process, 80–81; social impact of, 13, 19
Intergovernmental Panel on Climate Change (IPCC), 191n16, 213n3
International Commission on Large Dams (ICOLD), 195n14
International Rivers organization, 31, 197n30
Ipixuna Dam, 37
Irá (activist), 60–61, 63, 103, 104, 178
Iriri Dam, 37
ISA (Socio-environmental Institute), 65, 74–75, 113–14, 116, 134–35, 136
Itaipú Dam, 32–33, 193n4
Itzigsohn, José, 210n11

Jarine Dam, 37
Jirau Dam, 40
Johannes (federal official), 89, 97, 100, 106
Jonas (FORT representative), 74, 75, 82, 92–93, 110

Jose (fisher), 115–16, 135, 137, 154, 174, 175–76
Julia (MPA representative), 131–32, 133
justice, 19, 20; *See also* distributive justice; environmental justice

Kakraimoro Dam, 37
Kararão Dams, 37, 58
Keck, Margaret, 204n38, 209n1
Khagram, Sanjeev, 203n29
Khan, Mizan, 19
Krauss, Celene, 177
Kräutler, Bishop Dom Erwin, 53, 54, 56, 67, 201n16, 201n18

Lacerda, Paula, 204n31, 204n33
LaDuke, Winona, 177
Latin America: debt crisis, 58; democratization of, 22
law: democratic developmentalism and, 80–85; social activism and, 142, 149–52, 175; struggles for justice and, 174–75; *See also* rule of law
legal action: *vs.* activism, 151
legal agencies: claims making, 175; creative strategies of, 175–76; democracy and, 181–82; legitimacy of, 151, 153–54, 157, 173–74; mediating role of, 164–65, 173–74, 176; partnership between scientists and, 173; political conflict and, 156; power of, 175; social movements and, 142–43, 149; structural barriers in, 152, 155–56, 175; support of marginalized communities, 174–75
liberation theology, 47, 52–53, 59, 60, 201n13
licensing processes, 12, 42, 65, 80–81, 82, 85
Lopes, José Antônio Muniz, 57, 57
Lucas (director of federal energy programs), 106–7
Lula. *See* Silva, Luiz Inácio Lula da

Maathai, Wangari, 177
MAB (Movement of People Affected by Dams): at CGDEX meetings, 89; creation of, 36; dam construction and, 69–70; housing project and, 103, 105; local communities and, 70, 128, 161; opposition to big infrastructure projects, 69, 164; political influence, 69, 70
Malin, Stephanie, 16
management councils, 41

Mari (Government House representative), 131–32
Marxism, 59
McAllister, Lesley, 83, 175
McCormick, Sabrina, 173, 203n29
MDTX (Movement for the Development of the Transamazon and Xingu), 62, 68
mediation process, 23, 169, 173–74, 176; *See also* translation process
Médici, Emílio Garrastazu, 50–51
Melo da Silva, Antonia: anti-dam protests and, 46–47, 60, 64–65, 67–68, 70, 89–90, 111; background of, 46; on Basic Environmental Project (PBA), 82; criticism of CGDEX, 75, 103, 104, 105–6; legal services, 61, 113, 140; outreach efforts, 145–46, 149, 159; social activism, 47, 49, 52, 53–54, 55–56, 63, 68–69, 178
Mika (activist), 54–55, 60–61, 63
Mill, John Stuart, 21
Mische, Ann, 23
mitigation programs: community involvement in, 43; government's role in, 42–43, 76, 77, 78–79; importance of, 172; private entities and, 42, 44, 76–77; shared responsibility for, 76–77, 78–80
MMTA-CC (Movement of Women Workers of Altamira- Country and City), 59
Montero, Alfred, 202n24
Mosse, David, 157
Movement of Black Women, 75
Movement of Women Workers of Altamira, 178
Mozambique: hydropower in, 30
MPA (Ministry of Fisheries and Aquaculture), 123, 128–29, 131, 132, 133–34
MPF (Federal Public Prosecutors' Office), 82–84, 85, 136, 151; *See also* public prosecutors
MP-PA (Public Prosecutors' Office of the State of Pará), 82
MPST (Movement for the Survival on the Transamazon), 61, 68
multinational development banks (MDBs), 202n25

National Confederation of Fishers, 118
negotiation, 167, 168, 169
Nehru, Jawaharlal, 29
neoliberalism, 79

nongovernment organizations (NGOs), 31, 159, 160, 163
Norte Energia: Belo Monte Dam construction and, 27, 77, 82, 156; communication team, 4–5; community meetings, 2–3, 4–5, 9, 42, 43, 169; compensation program, 8, 77, 117, 149; creation of, 78, 82; criticism of, 135; headquarters of, 156; legal challenges of, 83, 152–53, 155, 156, 157; managing directors of, 2–3; mitigation efforts, 42, 44, 76–77, 117, 156–57; protest at headquarters of, 73; public distrust to, 149; research of dam effects, 117, 134; resettlement program, 16, 103, 122, 135, 145; riverine communities and, 117, 124–25, 128–29, 130–33, 135–38, 147, 172–73; Social Monitoring Forum of, 43

Operation Lava Jato (Car Wash), 12–13, 78, 180, 189n3
Our Common Future ("Brundtland Report"), 203n28

PAC (Growth Acceleration Program), 11, 40, 66
Pakistan: hydropower in, 30
palafitas neighborhoods, 43, 121–22
participatory democracy, 21–23, 198n38, 208n8, 210n10
PAS (Sustainable Amazon Plan), 41
Paulo (president of fishers' union), 117–19, 120, 121, 125, 209n4
PBA (Basic Environmental Project), 42, 82
PDRSX (Regional Sustainable Development Plan of the Xingu), 93
Petrobras, 189n3
Pimental dam, 171
PIN (National Integration Plan), 50
processo vivo, 96
protest activity, 1–2, 21, 22, 168
public defenders: Altamira's legal team and, 158; Belo Monte conflict and, 83, 84–85, 152; Belo Sun mining project and, 163, 164; government agencies and, 157–58, 175; institutional challenges, 143–44, 151–52; legal cases, 154–55, 168; legitimacy of, 155, 164–65; local offices of, 84; as mediators, 155; meetings with riverine people, 145–46, 150; meetings with urban residents, 140–42, 141, 144, 148–49; outreach strategies, 142–43, 144, 148, 149, 150, 151, 158–59; Pará office, 153; professional and activist identity of, 212n8; *vs.* public prosecutors, 84, 157; support of marginalized communities, 24, 150, 152; trust building, 143, 156; *See also* DPE (Public Defenders' Office of the State)
public prosecutors: legitimacy of, 164–65; outreach strategies, 142–43, 158, 175; partnership network, 210n17; *vs.* public defenders, 157; state-run agencies of, 175; trust building, 156; *See also* MPF (Federal Public Prosecutors' Office)

Rafael (representative of Norte Energia), 156–57
Randeria, Shalini, 77, 210n8
representative democracy, 207n1
resettlement program, 16, 103–5, 122, 134–35, 140, 145; *See also* urban resettlement communities
RIMA (Summary of the Environmental Impact), 81
Rio de Janeiro Earth Summit, 203n28
riverine people: children schooling, 144–45; civic engagement, 115; compensation to, 117, 137, 138, 145, 146–47, 148, 180; disagreements between Norte Energia and, 135–36; ethnicity of, 146; impact of dam construction on, 114–15, 134; legal claims of, 154–55; living conditions, 144, 148, 161; protests of, 114; public defenders and, 145–46, 147–48; resettlement of, 134, 140, 145; struggle for rights, 113–14, 138, 146–47, 150–51, 154, 172
Riverine People's Council, 113, 137–38, 153, 154
Riverine People's Dialogues, 135
Roberts, J. Timmons, 19
Robinson, Mark, 197n34
Roddick, Anita, 57
Rodrigo (fishers' leader), 121–22, 124–25, 135, 137, 153–54
Roosevelt, Franklin D., 29
Rousseff, Dilma Vana, 12, 40, 41, 105, 166
rubber industry, 48, 49, 200n4
RUCs. *See* urban resettlement communities
rule of law, 142

Samuels, David, 204n38
Santi, Thais: background of, 7, 114; Belo Monte negotiations and, 7–8; building of

institutional legitimacy, 153–54; inspection of dam-affected areas, 135; reputation of, 176; research on dam impact, 136–37; support for riverine communities, 83, 114, 134, 136, 152–54, 155, 157, 179
SBPC (Brazilian Society for the Advancement of Science), 136
Schmink, Marianne, 50
Senador José Porfírio, 161
Sergio (leader of fishing group), 93, 95, 137
Shepard, Peggy, 177
Shiva, Vandana, 177
Silva, Luiz Inácio Lula da: corruption scandal, 189n3; developmental rhetoric of, 71, 166; economic policy, 66–67; infrastructure projects, 11–12, 40, 41, 64; participatory initiatives, 42, 198n38; presidential election, 11, 60, 63, 67; visit to Altamira, 64, 65, 66, 70
Silva, Marcelo, 23
Silva, Maria, 59
Simone (leader of FVPP): background, 51–52, 54; at CGDEX meeting, 97; legal consultations, 61; position on dam construction, 68, 71, 111; social activism, 52, 54, 56, 65, 179
sites of acceptance, 16
sites of resistance, 16, 86
Snake River dam, 30
social movements: Catholic Church and, 46, 52–54, 55, 61; challenges of, 163–64; collaboration between, 65; growth of, 61–62; legal institutions and, 149; mediating role of, 210n10; politics and, 61; split within, 67–68; women in, 46–47, 176–79
social welfare, 11, 13
Sonia (member of CGDEX), 109–10, 111
Special Belo Monte Working Group, 84
Stang, Dorothy, 178
STTR (Rural Workers' Union), 107
Sun Yat-sen, 29
sustainable development, 203n30
symmetry, 190n8

Tapajós dam complex, 144
Three Gorges Dam, 29, 193n4
Tocantins River dam, 33
Tocqueville, Alexis de, 21

Tomas (public defender), 154–55, 159, 162
Transamazon Highway, 3; blockade of, 54, 56; construction of, 33, 50, 200n10; map of, 39
translation process, 23, 132–33; See also mediation process
Tucuruí Dam: energy capacity of, 33, 34, 196n23; environmental impact, 33–34, 196n22
Tuíra (Indigenous protestor), 57, 57
Turkey: hydropower in, 30

unequal exchange theory, 200n1
United States: hydropower in, 29
United States Environmental Protection Agency, 18
Universidade Federal do Pará (UFPA), 55
Urban Resettlement Collectives (RUCs), 44
urban resettlement communities, 122, 123, 159, 179
urban residents: meetings with public defenders, 140–42, 141, 148–49

Vargas, Getúlio, 32, 49
Volta Grande (Big Bend) region: access to, 48; Belo Monte Dam's impact on, 159–60; ecosystem, 117, 120–21, 209nn3–4; living conditions, 161; location of, 38; map of, 39; public meeting in, 159–64

Washington, Karen, 177
White, Gordon, 197n34
Wolford, Wendy, 210n10
women: environmental movement, 176, 177–79; knowledge and authority, 177; leadership roles, 176–77, 178, 179; organizational power, 178; social activism, 60–61, 177–78, 179; violence against, 59, 178
Wood, Charles, 50
Workers' Party: approach to governance, 41, 79; economic strategy, 197n34; formation of, 59; growth of, 63; infrastructure projects of, 11, 40, 65, 80; rise to power, 38, 59–60, 85, 167, 204n38; social movement and, 61, 109, 204n38; support base of, 36, 67–68, 69, 70
Working Group of Fishing, 137, 174
World Bank, 36, 37, 57–58, 67, 202n25, 203n29
World Commission on Dams, 193n1
World Conservation Union, 193n1

Xingu region: development of, 49, 89; social mobilization in, 47

Xingu River: boat navigation on, 2–3, 117, 118–19; dams on, 11, 15, 20–21, 36–37, 62, 197n31, 203n29; dead vegetation on, 160; *See also* Volta Grande (Big Bend) region

Xingu Vivo (Xingu Forever Alive Movement): Big Bend meeting, 159; CGDEX and, 75, 86, 111–12; criticism of, 125; foundation of, 46, 60; housing strategies, 103; leadership of, 89, 140; opposition to big infrastructure projects, 164; outreach efforts, 149; protest activities, 89, 111–12; publicity, 162; relations between MAB and, 69–70; resettlement programs, 103–5; split within, 64; support of fishers, 124, 134; women role in, 178

Yangtze River dam, 29

ABOUT THE AUTHOR

PETER KLEIN is an assistant professor of sociology and environmental and urban studies at Bard College. He earned his PhD in sociology from Brown University. In addition to publishing numerous articles, Klein is co-author of *The Civic Imagination: Making a Difference in American Political Life*.

Available titles in the Nature, Society, and Culture series

Diane C. Bates, *Superstorm Sandy: The Inevitable Destruction and Reconstruction of the Jersey Shore*

Soraya Boudia, Angela N. H. Creager, Scott Frickel, Emmanuel Henry, Nathalie Jas, Carsten Reinhardt, and Jody A. Roberts, *Residues: Thinking through Chemical Environments*

Elizabeth Cherry, *For the Birds: Protecting Wildlife through the Naturalist Gaze*

Cody Ferguson, *This Is Our Land: Grassroots Environmentalism in the Late Twentieth Century*

Albert Fu, *Risky Cities: The Physical and Fiscal Nature of Disaster Capitalism*

Shaun A. Golding, *Electric Mountains: Climate, Power, and Justice in an Energy Transition*

Aya H. Kimura and Abby Kinchy, *Science by the People: Participation, Power, and the Politics of Environmental Knowledge*

Peter Taylor Klein, *Flooded: Development, Democracy, and Brazil's Belo Monte Dam*

Anthony B. Ladd, ed., *Fractured Communities: Risk, Impacts, and Protest against Hydraulic Fracking in U.S. Shale Regions*

Stefano B. Longo, Rebecca Clausen, and Brett Clark, *The Tragedy of the Commodity: Oceans, Fisheries, and Aquaculture*

Stephanie A. Malin, *The Price of Nuclear Power: Uranium Communities and Environmental Justice*

Stephanie A. Malin and Meghan Elizabeth Kallman, *Building Something Better: Environmental Crises and the Promise of Community Change*

Kari Marie Norgaard, *Salmon and Acorns Feed Our People: Colonialism, Nature, and Social Action*

J. P. Sapinski, Holly Jean Buck, and Andreas Malm, eds., *Has It Come to This?: The Promises and Perils of Geoengineering on the Brink*

Chelsea Schelly, *Dwelling in Resistance: Living with Alternative Technologies in America*

Sara Shostak, *Back to the Roots: Memory, Inequality, and Urban Agriculture*

Diane Sicotte, *From Workshop to Waste Magnet: Environmental Inequality in the Philadelphia Region*

Sainath Suryanarayanan and Daniel Lee Kleinman, *Vanishing Bees: Science, Politics, and Honeybee Health*

Patricia Widener, *Toxic and Intoxicating Oil: Discovery, Resistance, and Justice in Aotearoa New Zealand*